Exploiting New Technologies in Animal Breeding

Genetic Developments

Exploiting New Technologies in Animal Breeding
Genetic Developments

EDITED BY

C. SMITH

AFRC Animal Breeding Research Organization,
West Mains Road,
Edinburgh, UK

J.W.B. KING

AFRC Animal Breeding Liaison Group,
West Mains Road,
Edinburgh, UK

and J.C. McKAY

AFRC Poultry Research Centre
Roslin, Midlothian, UK

Proceedings of a Seminar in the CEC Animal Husbandry Research Programme held in Edinburgh, UK, on 19–20 June 1985. Sponsored by the Commission of The European Communities, Directorate-General for Agriculture, Co-ordination of Agricultural Research.

PUBLISHED ON BEHALF OF THE COMMISSION
OF THE EUROPEAN COMMUNITIES BY
OXFORD UNIVERSITY PRESS · 1986

Oxford University Press, Walton Street, Oxford OX2 6DP
Oxford New York Toronto
Delhi Bombay Calcutta Madras Karachi
Petaling Jaya Singapore Hong Kong Tokyo
Nairobi Dar es Salaam Cape Town
Melbourne Auckland
and associated companies in
Beirut Berlin Ibadan Nicosia

Oxford is a trade mark of Oxford University Press

Published in the United States
by Oxford University Press, New York

Publication No. EUR 10053 of the
Commission of the European Communities,
Directorate-General Telecommunications, Information
Industries and Innovation, Luxembourg

LEGAL NOTICE
Neither the Commission of the European Communities nor
any person acting on behalf of the Commission is
responsible for the use which might be made of the
following information.

British Library Cataloguing in Publication Data

Exploiting new technologies in animal breeding genetic developments: proceedings of a
seminar in the CEC Animal Husbandry Research Programme held in Edinburgh, UK, on
19-20 June 1985.
1. Livestock—Breeding I. Smith, C. (Charles) II. King, J. W. B.
III. McKay, J. C. IV. CEC Animal Husbandry Research Programme
636.08'21   SF105
ISBN 0-19-854209-7

Library of Congress Cataloging-in-Publication Data
Exploiting new technologies in animal breeding.
1. Livestock—Breeding—Congresses. 2. Livestock—Genetic engineering—Congresses.
3. Embryo transplantation—Congresses. I. Smith, C. (Charles), 1932-
II. King, J. W. B III. McKay, J. C.
SF105.E97   1986 / 636.08'2   86-17964
ISBN 0-19-854209-7

Set by Colset Private Limited, Singapore
Printed in Great Britain by
Butler & Tanner Ltd, Frome and London

Preface

The objectives of the Seminar were to provide an up-to-date summary of current work with new technologies in animal breeding, and its applications, and to consider future possibilities in research and development. The Seminar served these functions well and the contributions by the 20 speakers were well received by the 50 participants. With half of the allocated time for presentations, the discussions were active and useful.

The participants were from nine EC countries, and from a wide range of disciplines and research topics, so there was a full interaction and exposure to different fields, knowledge, and interests. In some cases the disparity was rather wide, and the overlap was limited, but the exposure to different ideas, techniques and possibilities was stimulating and appreciated. It is hoped that the published papers and discussions will also stimulate readers in a similar way, and extend their interests.

The new technologies often open up new possibilities with current breeding and improvement methods, and increase and extend their application in practice. This is especially true for methods of reproductive manipulation, such as embryo transfer and sexing. The advantage is that these can further improve proven methods of breeding, to produce improved livestock balanced for economic performance in production traits.

The new techniques of genetic manipulation are impressive in their specificity, their power and their possibilities. They allow developments not previously possible and their scope seems immense. The scientific achievements are impressive and the speed of developments is very fast. No doubt there will be many problems such as in control, stability and in reproductive fitness, with transgenic animals, in their application to practical livestock improvement. Also considerable facilities will be needed, as with conventional methods, to breed and evaluate the lines produced and to establish their competitive position for commercial production. To this extent the old and new technologies must complement each other in their development.

The European nature of the meeting, made possible by CEC funding, raised the spirit of co-operation and common purpose in animal breeding research. The value of the research and development in livestock improvement to society was expressed, with need for co-operation in sharing research funds on a European dimension. These are interesting and changing times in science and technology and this Seminar, and others like it, are necessary to understand the trend and direct it for the common good.

Edinburgh   C.S.
September 1985

Contents

Contributors

A.L. ARCHIBALD

AFRC Animal Breeding Research Organization, West Mains Road, Edinburgh EH9 3JQ, UK

J.O. BISHOP

Department of Genetics, University of Edinburgh, EH9 3JN, UK

P. BOOMAN

Research Institute of Animal Production, 'Schoonoord', PO Box 501, 3700 AM Zeist, Netherlands

G. BREM

Institute of Animal Breeding, Ludwig-Maximilians-University, Veterinärstrasse 13, 8000 Munich 22, Federal Republic of Germany

I. BURGUETE

Department of Genetics, Faculty of Veterinary Sciences, University of Leon, Leon, Spain

F. CAPUANO

Animal Production Institute, University of Naples, Faculty of Agriculture, 80055 Portici, Naples, Italy

L.G. CHRISTENSEN

National Institute of Animal Science, Research Center Foulum, PO Box 39, 8833 Ørum Sdrl., Denmark

A.J. CLARK

AFRC Animal Breeding Research Organization, West Mains Road, Edinburgh EH9 3JQ, UK

D.DI BERARDINO

Animal Production Institute, University of Naples, Faculty of Agriculture, 80055 Portici, Naples, Italy

M. FÖRSTER

Institute of Animal Genetics, Technical University of Munich, D-8050 Freising-Weihenstephan, Federal Republic of Germany

F. GANNON

Department of Microbiology, University College Galway, Co. Galway, Ireland

J.P. HANRAHAN

The Agricultural Institute, Belclare, Tuam, Co. Galway, Ireland

R. HANSET

Faculté de Médecine Vétérinaire (U.Lg), Rue de Vétérinaires 45, B-1070 Brussels, Belgium

L. IANNUZZI

CNR Institute on Adaptation of Cattle and Buffalo to the Southern Italy Environment (IABBAM), 80147 Ponticelli, Naples, Italy

E. KANIS

Department of Animal Breeding, Agricultural University, PO Box 338, 6700 AH Wageningen, The Netherlands

R.F. LATHE

Institut de Chimie Biologique, 11 Rue Humann, 67085 Strasbourg, France

C. LEGAULT

Institut National de la Recherche Agronomique, Station de Génétique quantative et appliquée, 78350 Jouy-en-Josas, France

T. LIBORIUSSEN

National Institute of Animal Science, Research Center Foulum, PO Box 39, 8833 Ørum Sdrl., Denmark

M.B. LIOI

Animal Production Institute, University of Naples, Faculty of Agriculture, 80055 Portici, Naples, Italy

R.H. LOVELL-BADGE

MRC Mammalian Development Unit, Wolfson House, University College London, 4 Stephenson Way, London NW1 2HE, UK

P. LØVENDAHL

National Institute of Animal Science, Research Center Foulum, PO Box 39, 8833 Ørum Sdrl., Denmark

T.F.C. MACKAY

Institute of Animal Genetics, West Mains Road, Edinburgh EH9 3JN, UK

J. R. MANN
Birth Defects Research Institute, Royal Childrens Hospital, Flemington Road, Parkville 3052, Victoria, Australia

D. MATASSINO
Animal Production Institute, University of Naples, Faculty of Agriculture, 80055 Portici, Naples, Italy

J. C. MERCIER
Laboratoire de Génétique Biochimique, Unité de Recherches en Biotechnologie, Centre National de Recherches Zootechniques, Institut National de Recherche Agronomique, 780350 Jouy-en-Josas, France

A. J. MOORE
AFRC Animal Breeding Research Organization, West Mains Road, Edinburgh EH9 3JQ, UK

E. MÜLLER
Institute of Animal Husbandry and Breeding, University of Hohenheim, Postfach 700562/470, D-7000 Stuttgart 70, Federal Republic of Germany

H. SANG
AFRC Poultry Research Centre, Roslin, Midlothian EH25 9PS, UK

K. SEJRSEN
National Institute of Animal Science, Research Center Foulum, PO Box 39, 8833 Ørum Sdrl., Denmark

P. SELLIER
Institut National de la Recherche Agronomique, Station de Génétique quantative et appliquée, 78350 Jouy-en-Josas, France

P. SIMONS
AFRC Animal Breeding Research Organization, West Mains Road, Edinburgh, EH9 3JQ, UK

C. SMITH
AFRC Animal Breeding Research Organization, West Mains Road, Edinburgh EJ9 3JQ, UK

P. STAM
Department of Genetics, Agricultural University, Foulkesweg 53, 6703 BM Wageningen, The Netherlands

St. C.S. TAYLOR

AFRC Animal Breeding Research Organization, West Mains Road, Edinburgh EH9 3JQ, UK

R.B. THIESSEN

AFRC Animal Breeding Research Organization, West Mains Road, Edinburgh EH9 3JQ, UK

I. WILMUT

AFRC Animal Breeding Research Organization, West Mains Road, Edinburgh EH9 3JQ, UK

1

Cytogenetic analysis of embryos

D. Di Berardino, M.B. Lioi, F. Capuano, L. Iannuzzi,
I. Burguete, and D. Matassino

Abstract

The preliminary results of a cytogenetic analysis on day-6 rabbit embryos are reported, in relation to: (a) RBA-banding, (b) nucleolus organizer regions (Ag-NORs), and (c) sister chromatid exchanges (SCEs). The investigation demonstrates that cytogenetic analysis on pre-implantation embryos can be performed with nearly the same degree of accuracy and resolution normally achieved in somatic cells. The usefulness of the RBA-banding procedure for sexing the embryos and for detecting embryos which are carriers of chromosomal abnormalities is stressed, as well as the opportunity to study the Ag-NORs distribution, the rate of SCEs, and chromosome breakages, for a more accurate evaluation of embryos. Cytogenetic analysis can also provide useful information on processes such as (a) inactivation of the X-chromosome in females, and (b) replication pattern of the R-bands in undifferentiated cells, to enable better understanding of some of the mechanisms involved in gene activation or inactivation during cell differentiation.

Introduction

Metaphase chromosomes of embryos were first visualized in pig (Bomsel-Helmreich 1961; McFeely 1966) and in mouse (Tarkowsky 1966); subsequent studies concerned rabbit (Shaver and Carr 1967), cattle (McFeely and Rajakoski 1968), sheep (Long 1974), and other domestic species. Cytogenetic analysis of embryos utilized mainly the conventional Giemsa staining for sexing the embryo and detecting chromosomal abnormalities. This procedure, however, has serious limitations because it allows the identification of the sex chromosomes only in those species whose gonosomes are morphologically different from the autosomes as in cattle. Furthermore, only large chromosomal abnormalities can be detected, whereas those involving small segments escape detection. These considerations indicate that for a more accurate cytogenetic analysis, it is necessary to adopt banding techniques. Other cytogenetic methods, such as the nucleolus organizer regions (Ag-NORs), the sister chromatid exchanges (SCEs), and chromosomal breaks could provide additional information, useful for a better evaluation of the embryos. The present paper demonstrates that cytogenetic analysis of

1

pre-implantation embryos can be performed with nearly the same degree of accuracy and resolution normally achievable in somatic cells.

Materials and methods

Females from the white New Zealand breed of rabbit were slaughtered 6 days after mating. The uterine horns were washed with RPMI 1640 medium (Gibco) supplemented with 20% fetal calf serum (FCS, Gibco), heat inactivated, and the blastocysts collected in watch glasses, washed twice in the same medium and incubated at 37°C in CO_2, according to the following requirements.

RBA-banding

The culture medium was enriched with bromo-deoxyuridine (BrdU, Sigma) at a final concentration of 10 μg/ml and the incubation lasted 6–7 h in the dark. Three hours before the end of the culture, colcemid (Gibco) was added at a final concentration of 0.2 μg/ml.

Nucleolus organizer regions (Ag-NORs)

The blastocysts were incubated for 4–6 h in the same medium (RPMI 1640 + 20 per cent FCS) in the presence of colcemid (0.2 μg/ml, final concentration).

Sister chromatid exchanges (SCEs)

The blastocysts were cultured in the dark for 24 h in culture medium enriched with BrdU (10 μg/ml, final concentration). Three hours before the end of the culture, colcemid was added at a final concentration of 0.2 μg/ml.

At the end of the incubation time, the blastocysts were washed in RPMI medium with no FCS, treated with trypsin (Difco) (0.2 per cent in PBS, pH = 8.0) for 10 min at 37°C to digest off the zona, washed again in fresh RPMI medium, and hypotonized first with sodium citrate (0.8 per cent) for 5 min and then with potassium chloride (0.075M) for a further 5 min, in Eppendorf tubes. The hypotonic treatment was stopped by the gentle addition of fixative (3:1 methanol-acetic acid) until complete substitution of the potassium chloride occurred. The last fixation was prolonged overnight at 4°C. At the end of the fixation treatment, each blastocyst was located at the centre of a microscope slide with a drop of fixative, gently dispersed on the glass by mild pipetting, and air dried.

Staining procedure

The cytological preparations suitable for RBA-banding were stained with acridine orange (0.1 per cent) in phosphate buffer (pH = 7.0) for 10 min, washed in tap water, mounted in the same buffer, and sealed with paraffin.

The same procedure was applied to the slides for the SCE. The visualization of the nucleolus organizer regions (Ag-NORs) was achieved by floating the slides with a 50 per cent solution (w/v) of Ag-NO$_3$ in distilled water, covering with a coverslip, locating the slide in a moist chamber, and incubating at 37°C for 24–36 h.

Results and discussion

RBA-banding

Figure 1.1 shows the RBA-banding pattern in two metaphase plates of the same male rabbit embryo. The cell in Fig. 1.1A is aneuploid ($2n = 41$), lacking the X chromosome and two autosomes. The Y chromosome (arrowed) is easily identifiable, being almost entirely heterochromatic and pale, as described by Hagelthorn and Gustavsson (1979). The cell shown in Fig. 1.1B is normally diploid ($2n = 44$), and even if the chromosomes are very contracted, as usually occurs in embryo preparations, the Y chromosome is always easily identifiable (arrow). Figure 1.2 shows the RBA-banding pattern in two metaphase plates of the same female rabbit embryo. The cells shown in Figs. 1.2A and 1.2B are both diploid; the inactive X chromosome is easily identifiable (arrows) being almost entirely heterochromatic and pale for most of its length. From the cell shown in Fig. 1.2A, the RBA-banded karyotype shown in Fig. 1.3 was obtained, according to the Reading nomenclature (Ford *et al.* 1980). As can be seen, the quality of the RBA-banding pattern is such that the embryo can be sexed and karyotyped precisely.

The RBA-banding procedure was recently proposed by Popescu (1983) for cattle embryos, but the author utilized cytological preparations obtained from cells removed by biopsy and grown on a collagen layer for 5–6 days. The method reported in the present paper is faster, being particularly suitable for basic cytogenetic studies, but has the disadvantage that the embryo is lost. Some studies where the technique presented herein might be applied include; (a) the inactivation of the X chromosome in females of the different domestic species, and (b) the replication pattern of the R-bands in undifferentiated (or little differentiated) cells such as the blastomeres and in highly specialized cells, such as lymphocytes or fibroblasts.

Nucleolus organizer regions (Ag-NORs)

Figure 1.4A shows a metaphase plate of an embryo with six nucleolar chromosomes with telomeric NORs, while Fig. 1.4B shows a nucleolar association between two chromosomes, and Fig. 1.4C shows an interphase nucleus with five clusters of silver nitrate material, corresponding to five nucleoli and to ten chromosomes carrying the NORs. These preliminary observations seem to confirm the presence, in the rabbit, of five nucleoli and a number of NORs per metaphase variable from 3 to 6 (Martin *et al.* 1978), never reaching 10. It seems, therefore, that at this stage of embryonic

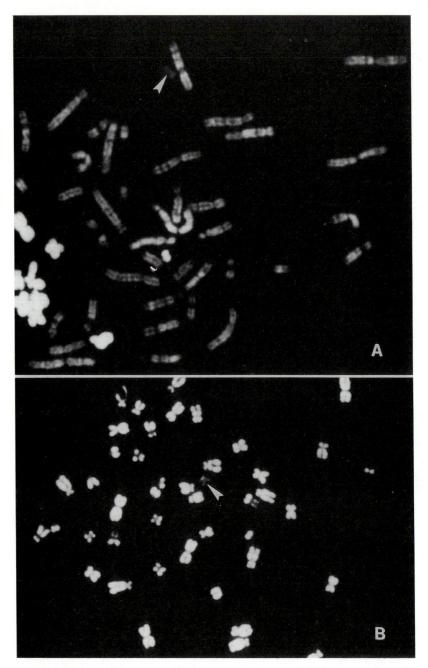

Fig. 1.1.   RBA-banding pattern (A) in a prometaphase and (B) in a metaphase plate of the same day-6 male rabbit embryo. Arrows indicate the Y chromosome.

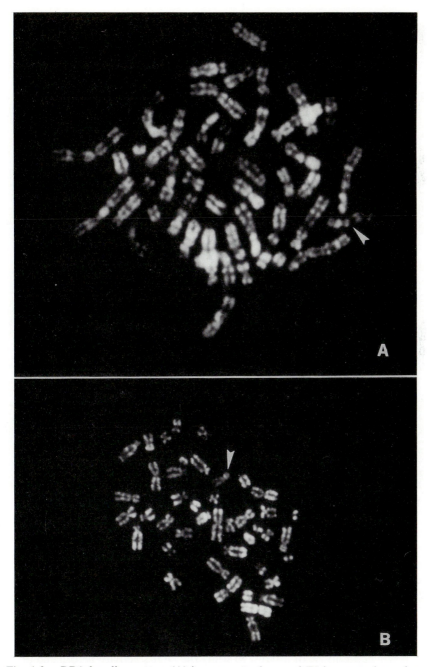

Fig. 1.2. RBA-banding pattern (A) in a prometaphase and (B) in a metaphase plate of the same day-6 female rabbit embryo. Arrows indicate the inactive X chromosome.

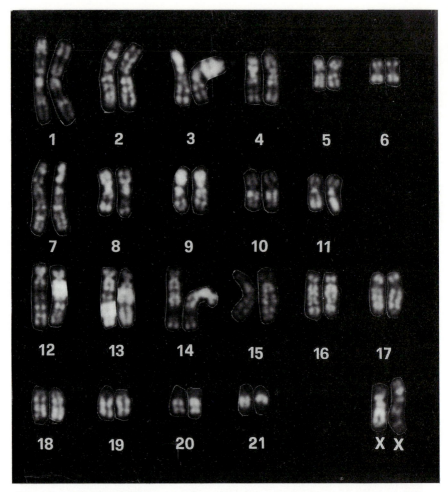

Fig. 1.3.   RBA-banded karyotype from the same prometaphase plate shown in Plate 2A, arranged according to the Reading nomenclature.

development, the sites for the production of 18S + 28S ribosomal RNA cannot be visualized, or are not active as in the somatic cells of adult subjects (Goodpasture and Bloom 1975; Howell *et al.* 1975). The data also indicate that the rRNA activity is already 'on' in day-6 rabbit embryos, i.e. before gastrulation, while in Amphibia, for example, transcription of rRNA is known to start during gastrulation (Balinsky 1975).

It will certainly be of interest to study the NORs activity before and after this stage of embryonic development among different domestic species.

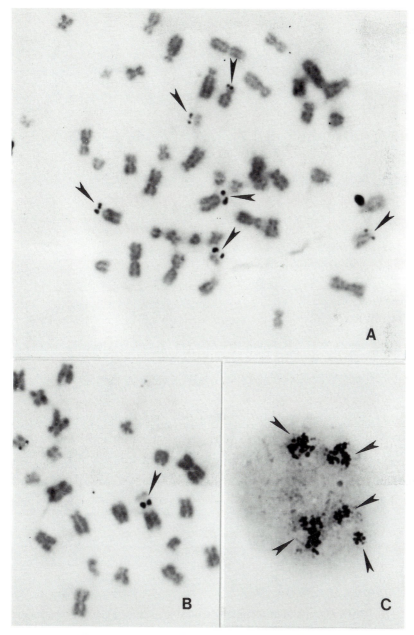

Fig. 1.4. (A) Silver-stained metaphase plate of day-6 rabbit embryo showing 6 telomeric NORs. (B) Nucleolar association between two chromosomes. (C) Inter-phase nucleus showing five clusters of silver material, corresponding to five nucleoli.

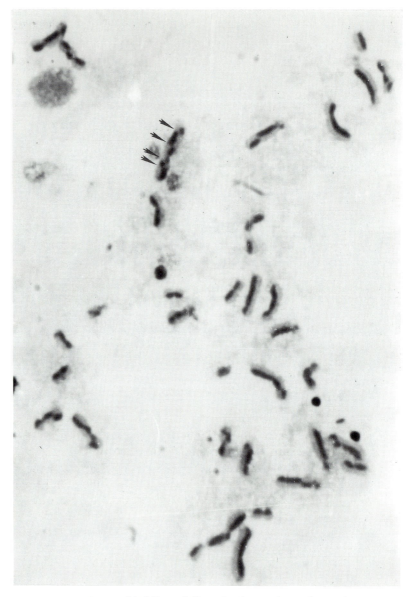

Fig. 1.5. Sister chromatid differentially stained metaphase plate (Giemsa counter-stained) of a day-6 rabbit embryo showing 23 exchanges, four of which are located on one large biarmed chromosome (arrows).

Sister chromatid exchanges (SCEs)

Figure 1.5 shows a partial metaphase plate ($2n = 35$) with 23 exchanges; in particular one of the chromosomes shows four exchanges, two in the short arm and two in the long one. The BrdU concentration was 10 μg/ml, the same normally used for SCE studies on lymphocytes or fibroblasts. In our experiments, this concentration induced a drastic reduction in mitotic index and a rate of SCE per cell higher than the normal one. It is possible that these effects are due to the BrdU concentration, which was relatively higher when compared to the cell density (Stetka and Carrano 1977), or to a higher sensitivity of the embryonic cells to BrdU, which is known to be a strong mutagenic agent. Studies are underway in order to optimize BrdU concentration with the different stages of embryonic development.

The SCE test alone, or combined with the estimation of chromosome or chromatid breaks (Di Berardino *et al.*, 1983) is, at the moment, the most sensitive indicator of the cytogenetic damage induced by mutagens, and provides a measure of the degree of stability (or instability) of the genome. Applied to embryos, this test could provide useful information for detecting genetic disorders associated with deficiency in the DNA repair systems, such as the Bloom syndrome in humans (Chaganti *et al.* 1974), for testing the effects of environmental mutagens and carcinogens, and for selecting 'unstable' genotypes. The embryo, with regard to this, can be considered the best *in vivo* system.

Conclusions

Cytogenetic analysis of embryos, so far, has been focused on sexing and detection of chromosomal abnormalities by using conventional Giemsa staining. As far as we know, a more accurate cytogenetic study including the RBA-banding, the NORs visualization, the SCE test combined with the estimation of chromosome/chromatid breakages, has never been undertaken. The present paper demonstrates the opportunity to use the RBA-banding procedure for 'sexing' the embryos as well as for detecting embryos with abnormal karyotypes. Other banding techniques, such as G- or Q-banding, could also be utilized but, in our experience in domestic species, the RBA-banding has proved to be faster, more reliable, and more sensitive (Di Berardino and Iannuzzi 1982, 1984). The cytogenetic sexing of embryos could also be performed by using the C-banding technique (constitutive heterochromatin), as recently demonstrated by Murer-Orlando *et al.* (1982) in horse, but this method does not allow detection of chromosomal abnormalities. The RBA-banding method reported in the present paper was applied to the embryo *in toto* but we believe that it could also be applied to a fragment of the blastocyst, according to the procedure reported by Hare *et al.*

(1980). Alternatively, the method proposed by Popescu (1983) on a collagen layer is recommended.

The usefulness of a precise and accurate cytogenetic analysis is practical and scientific. Four out of ten of the embryos collected after superovulation in cattle are classified as 'non-vital' and eliminated; of the six remaining embryos, only three will result in pregnancy (Seidel 1981; Sreenan 1983). The causes of this low efficiency are still unknown; the available data, however, indicate that up to 60 per cent of the abortions may be due to chromosomal abnormalities, especially when superovulatory techniques are used (Fujimoto *et al.* 1974; King 1983). Also, the embryos which are normally transplanted are not checked for chromosomal abnormalities, and may well include a certain proportion of aberrations. As is well known, balanced structural chromosomal abnormalities are generally compatible with life, but have deleterious effects on fertility, as demonstrated by Gustavsson (1969, 1979) in cattle affected by the Robertsonian translocation 1/29 in a heterozygous state. On the other hand, unbalanced aberrations such as monosomy, nullisomy, and double trisomy are generally lethals, and the embryo dies early during embryogenesis.

In addition to the practical advantages, cytogenetic analysis can be a powerful tool for a better understanding of the cytology of embryos. Studies on the rRNA activity, on the SCE response and on chromosome breakages might prove useful for the evaluation of the embryos and their vitality. Furthermore, the embryo can be considered as the best *in vivo* system for evaluating the cytogenetic damage from environmental mutagens and carcinogens. Other cytological aspects, such as the process of inactivation of the X chromosomes in females of the different domestic species and, the replication pattern of R-bands in undifferentiated and specialized cells can also prove useful for understanding some of the mechanisms involved in gene activation or inactivation during cell differentiation.

References

Balinsky, B. I. (1975). *An introduction to embryology* (4th edn). W. B. Saunders Co., Philadelphia.

Bomsel-Helmreich, O. (1961). Heteroploidie experimentale chez la truie. *Proc. IV Int. Congr. Anim. Reprod. Insem. Artif., La Haye* **98**, 1–4.

Chaganti, R. S. K., Schonberg, S., and German, J. (1974). A manyfold increase in sister chromatid exchanges in Bloom's syndrome lymphocytes. *Proc. Nat. Acad. Sci.* **71**, 4508.

Di. Berardino, D. and Iannuzzi, L. (1982). Detailed description of R-banded bovine chromosomes. *J. Hered.* **73**, 434–8.

—— and —— (1984). Detailed description of RBA-banded chromosomes of river buffalo (*Bubalus bubalis* L.). *Génét. Sél. Evol.,* **16**, 249–60.

——, ——, Fregola, A., and Matassino, D. (1983). Chromosome instability in a calf affected by congenital malformation. *Vet. Rec.* **112**, 429–32.

Ford, C. E., Pollock, D. L., and Gustavsson, I. (1980). Proc. of the 1st Int. Conf. for the Standardization of banded karyotypes of domestic animals. *Hereditas* **92**, 145–62.

Fregola, A., and Matassino, D. (1983). Chromosome instability in a calf affected by congenital malformation. *Vet. Rec.* **112**, 429–32.

Fujimoto, S., Pahlavan, N., and Duklow, W. R. (1974). Chromosome abnormalities in rabbit preimplantation blastocysts induced by superovulation. *J. Repr. Fert.* **40**, 177–81.

Goodpasture, C. and Bloom, S. E. (1975). Visualization of nucleolus organizer regions in mammalian chromosomes using silver staining. *Chromosoma* **53**, 37–50.

Gustavsson, I. (1969). Cytogenetics, distribution and phenotypic effects of a translocation in Swedish cattle. *Hereditas* **63**, 68–169.

—— (1979). Distribution and effects of the 1/29 Robertsonian Translocation in cattle. *J. Dairy Sci.* **62**, 825–35.

Hagelthorn, M. and Gustavsson, I. (1979). Identification by banding techniques of the chromosomes of the domestic rabbit (*Oryctolagus cuniculus* L.). *Hereditas* **90**, 269–79.

Hare, W. C. D., Singh, E. L., Betteridge, K. J., Eaglesome, M. D., Randhal G. C. B., Mitchell, D., Bilton, R. J., and Trowson, A. O. (1980). Chromosomal analysis of 159 bovine embryos collected 12 to 18 days after estrus. *Can. J. Genet. Cytol.* **22**, 615–26.

Howell, W. M., Denton, T. E., and Diamond, J. R. (1975). Differential staining of the satellite regions of human acrocentric chromosomes. *Experientia* **31**, 260–2.

King, W. A. (1983). Cytogenetics of embryos: application to livestock industry. *Proc. 2nd Symp. Adv. Top. Anim. Reprod., 8–11 August, Jaboticabal, Brasil.*

Long, S. 1974. Fertility of sheep with a Robertsonian Translocation. *Vet. Rec.* **94**, 161–2.

Martin, P. A. De Leon, Petrowsky, D. L., and Fleming, E. M. (1978). Nucleolar Organizer Regions in the rabbit (*Oryctolagus cuniculus* L.) as shown by silver staining. *Can. J. Genet. Cytol.* **20**, 377–82.

McFeely, R. A. 1966. A direct method for the display of chromosomes from early pig embryos. *J. Reprod. Fert.* **11**, 161–3.

—— and Rajakosky, E. (1968). Chromosome studies on early embryos of the cow. *Proc. VI Congr. Anim. Reprod., Paris,* 905–7.

Murer-Orlando, M., Betteridge, K. J. and Richer, C. L. (1982). Cytogenetic sex determination in cultured cells of pre-attachment horse embryos. *5th Eur. Coll. on Cytogenetics of Dom. Anim., 7–11 June, Milano, Gargnano.*

Popescu, C. P. (1983). Micromanipulation in embryo sexing. *Proc. 2nd Symp. Adv. Top. Anim. Reprod., 8–11 August, Jaboticabal, Brasil.*

Seidel, G. E. 1981. Superovulation and embryo transfer in cattle. *Science* **211**, 351–8.

Shaver, E. L. and Carr, D. H. (1967). Chromosome abnormalities in rabbit blastocysts following delayed fertilization. *J. Reprod. Fert.* **14**, 415.

Sreenan, J. M. (1983). Embryo Transfer procedure and its use as a research technique. *Vet. Rec.,* **112**, 494–500.

Stetka, D. G. and Carrano, A. V. (1977). The interaction of H33258 and BrdU substituted DNA in the formation of Sister Chromatid Exchanges. *Chromosoma (Berl.)* **63**, 21–31.

Tarkowsky, A. K. (1966). An air drying method for chromosome preparations from mouse eggs. *Cytogenet.*, **5**, 394–400.

Discussion

Förster asked if specific staining shows any variation in rDNA in the embryos, but Di Berardino replied that he had no evidence for such variation within embryos. It would, however, be interesting to establish when rDNA activity begins in development. Booman wondered if embryos were classified according to quality. Di Berardino replied that they were, but as yet there was insufficient data to establish whether there is any correlation between quality classification and the incidence of chromosomal abnormalities. This prompted King to ask if more cytogenetic abnormalities exist in ova from superovulation? Di Berardino replied that there are certainly such claims in the literature for animals and man. More data are needed, but the dose of PMSG may be crucial. Land wondered what proportion of true abnormalities can be detected by these methods? Di Berardino said that the whole range of major chromosomal changes (deletions, translocations, inversions, etc.) are detected, but what proportion of phenotypic abnormalities this constitutes is not clear in farm species. In man, 60% of spontaneous abortion is associated with gross cytogenetic abnormality. We require better staining and banding techniques especially in cattle before good data can be collected. However, Booman commented that in cattle embryos classified as 'good quality', the incidence of chromosomal abnormalities seems quite low. Di Berardino agreed, but pointed out that the poorer quality embryos may be those with chromosomal abnormalities, while other chromosomal abnormalities may be perfectly viable but cause problems of fertility. Baronos wondered if farmers are correct to resist the use of superovulation in the belief that it gives abnormal offspring. Di Berardino felt that a high level of chromosomal abnormalities may be important, but we need much more data in all species to measure the incidence.

2

Control of sex ratio by sexing sperm and embryos

P. Booman

Introduction

The sex of a mammal is determined at fertilization and depends on whether the X-bearing haploid ovum is fertilized by an X- or Y-bearing haploid spermatozoon. Thus, sex could be predetermined if X and Y spermatozoa were separated before insemination, and this would undoubtedly be the ideal method of controlling sex ratios.

Since the separation of X- and Y-bearing spermatozoa, as we will see, is not an immediate prospect at the moment, the ability to determine the sex of embryos prior to transfer in conjunction with embryo transfer techniques would be of great benefit to animal producers. There are two basic methods of sexing embryos. The first involves examination of diploid cells at a suitable stage after fertilization to see whether their sex chromosomes are XX (female) or XY (male). This procedure is called karyotyping. The second method of sexing is based on the detection of an antigen that is peculiar to cells containing a Y chromosome.

Separation of sperm

The possibility of separating sperm bearing the X from those bearing the Y chromosome has since long interested scientific investigators, physicians and animal breeders. Numerous reports have been presented supposedly demonstrating that such a separation is possible based on physical methods. These include sedimentation, counterstream centrifugation, or electrophoresis of the cells under a variety of conditions, and the use of systems that depend on differential migration of spermatozoa in colloidal media. Claims of success in recent years, coupled with advances in biological knowledge and technology, have resulted in three broad-based symposia to evaluate the present state and future prospects (Kiddy and Hafs 1971; Amann and Seidel 1982; Gledhill and Garner 1983). At none of the meetings was there general acceptance of the success of any method. The recurrent theme in this area of investigation has been the inability to reproduce results. The limiting factors in the separation of X- and Y-bearing mammalian sperm by physical means are the relative magnitudes of the variations between the two classes of cells and the variations within each class.

Research on techniques designed to separate sperm cells has been hampered by the lack of laboratory tests to evaluate the degree of separation. In 1969 in man, prospective investigation of the effect of various treatments on the X/Y sperm ratio in the ejaculate became possible with the discovery of a method for 'Y' body staining (Zech 1969). However, it is generally agreed that in domestic animals the Y-chromosome does not fluoresce (Pearson *et al.* 1971; Ericsson *et al.* 1973). So until recently, in farm animals the effectivity of any particular separation had to be judged by insemination trials to determine the sex ratio of progeny. Although insemination of superovulated females with supposedly enriched semen followed by embryo sexing might be another experimental approach, in 1982 Pinkel *et al.* described a very useful method to evaluate claims of enrichment. They reported a technique for measuring the nuclear DNA content by flow cytometry that allows determination of the proportion of X and Y chromosome-bearing spermatozoa. The process has been adapted for spermatozoa of domestic animals (Garner *et al.* 1983) and can be used to assess the effectiveness of methods of sperm separation much faster and cheaper than determining the sex ratio of embryos or live-born animals. For example, samples of bovine semen processed for enrichment with a variety of techniques turned out to have equal proportions of X and Y sperm when analysed with flow cytometry (Pinkel *et al.* 1983). At the present time, sorting sperm cells by flow cytometry is destructive and even if a marker of the chromosome constitution of viable sperm is found, flow cytometry cannot provide a practical answer for commercial application. The number of sperm required for each insemination is in the millions and is beyond what flow sorting can supply. However, flow cytometry can be of great importance for facilitating development of new techniques for sexing spermatozoa.

Most approaches to sperm separation have been primarily based on real or assumed physical differences between cells containing an X- or a Y-chromosome, rather than on specific haploid expression of genes. Provided that some of these gene products present themselves on the sperm plasma membrane, specific antibodies could be directed against either X- or Y-bearing sperm. Aside from the male specific H-Y antigen (Eichwald and Silmser 1955) several X-linked plasma membrane antigens are known (see for references, Ohno 1982). Unfortunately, according to Ohno (1982) there is little hope of utilizing gene products of the X-chromosome that might be expressed on the sperm plasma membrane as a means for sexing sperm. The entire X chromosome apparently remains dormant during spermatogenesis and spermiogenesis. Nevertheless, Koo *et al.* (1973) visually located the male specific H-Y antigen on the plasma membrane of sperm by electron-microscopy and spermatozoa are known to be excellent cytotoxic targets for specific antibodies directed against the H-Y plasma membrane antigen (Goldberg *et al.* 1971). It is clear, however, that expression of the H-Y antigen requires the Y-linked gene as well as the X-linked one (Ohno 1979; Wachtel and Ohno 1980). Ohno (1982) supposed that the H-Y antigen present in abundance on the sperm plasma membrane was actually contributed by Sertoli cells and not synthesized by germ cells with a dormant X chromosome. Thus, use of H-Y antigen as a means for selection of X or Y chromosome bearing sperm appears to be an unlikely prospect.

In this respect there are some contradictory indications. Bryant (1980) reported a large and significant change in the sex ratios when mice were inseminated with sperm passed through an immunoabsorbent column containing H-Y antibodies. Two populations of sperm cells were presumably obtained, one containing Y-bearing sperm and the other containing X-bearing sperm. Bryant (1980) reported sex ratios of 92 per cent male offspring when the 'Y' fraction was used and 96 per cent female offspring when the 'X' fraction was used. Similarly, Zavos (1983) reported that intravaginal administration of H-Y antisera to rabbits resulted in a significant increase in the number of female offspring (74 per cent). These results seem to indicate that the sperm cell surface reflects to some degree the haploid constitution of the sperm. However, other attempts to control the sex ratio by treating sperm with H-Y antibodies have met little success. Hoppe and Koo (1984), for example, treated spermatozoa with monoclonal H-Y antibody, but they failed to influence sex ratios. Furthermore, Pinkel *et al.* (1983) reported that sperm separated on the basis of their H-Y antigen constitution did not differ in their X:Y ratio. Thus, although the results obtained by Bryant (1980) and Zavos (1983) remain inexplicable, immunoselection with H-Y antibodies appears to have very little, if any, practical application.

Sexing embryos

Sex chromosomal analysis

Day-12 to day-15 bovine and ovine embryos have been successfully sexed before transfer by chromosome analysis on trophoblast biopsies. Betteridge *et al.* (1981) summarized results from five laboratories using this technique. They reported that 68 per cent of day-12 to day-15 bovine embryos can be sexed and that a 33 per cent pregnancy rate results from transfer of such embryos. In a recent review, King (1984) enumerated the limiting factors with the technique of trophoblast biopsy. These are the relative numbers of cells in metaphase, the quality of those metaphases, and the time taken for arriving at a diagnosis. Hare *et al.* (1976) found that two experienced cytogenetic technicians take about 5 h to process 12 to 15 embryos. Since day-12 to day-15 embryos cannot be stored for longer than a few hours, this makes it difficult to obtain the results and to perform the transfer on the same day.

Chromosome analysis has also been used on cells removed from unhatched, 6- and 7-day bovine embryos at transfer with subsequent birth of sexed calves (Moustafa *et al.* 1978; Schneider and Hahn 1979). The day-6 to day-7 bovine embryo can readily be collected, frozen, and transferred non-surgically. For sexing, small biopsies (10–20 cells) are collected by micro-manipulation. The technique of sexing bovine embryos on day 6 or day 7 using those biopsies is hampered by the necessity to work with an extremely small number of cells. This means that a very high mitotic index (over 10%) is necessary to provide sufficient cells in metaphase. This is generally believed to prevent effective use on a routine basis (King 1984).

In his review, King (1984) mentioned that the development of techniques for bisecting embryos of domestic animals, particularly bovin embryos, on days 6 or 7 has raised the possibility of sacrificing one of the half-embryos for chromosome analysis. Several groups have been exploring the possibility of sexing by this approach (see, for references, King 1984). Similarly, the limiting factors appear to be the number and the quality of the metaphases in the preparation.

Of the cytological methods that have been used to sex embryos, only trophoblast biopsies on day 12 to day 15 and embryo bisection on day 6 or 7 are considered to be practical at present. The first one excludes the possibility of long-term storage, while the second one requires additional expense and expertise for micromanipulation. Both methods require trained cytogeneticists, are time consuming, and affect the viability of the embryo. For these reasons it is unlikely that sex chromosomal analysis will be used for commercial embryo sexing. However, King (1984) concluded that these methods are useful for confirming results of alternative means of sexing embryos, since cytogenetic analysis determines the genetic sex of the embryo with little risk of false diagnosis.

Detection of H-Y antigen

A possible alternative method of sexing embryos is the serological demonstration of the H-Y antigen. The H-Y antigen was first detected in 1955, when it was observed that female mice rejected skin grafts from male syngenic mice (Eichwald and Silmser 1955) whereas skin grafts exchanged among the other sex combinations were accepted. Rejection in this case is due to H-Y (histocompatibility-Y) antigen, present in cells of the male, but not in those of the female (Billingham and Silvers 1960). Serum from females that have rejected male skin grafts contains antibodies (H-Y antibodies) that are able to identify male cells in serological systems. The molecule conferring H-Y antigenicity is phylogenetically conservative. H-Y antibodies of the mouse and rat have been used to identify XY-cells in some seventy species from all classes of vertebrates. Among the mammals, H-Y is present in males of cattle, sheep, goat, horse, pig, and human (see, for references, Wachtel

1983). Regarding its functional aspects, H-Y antigen has been suspected of taking part in the process of primary sex determination (Wachtel 1983, to which the reader is referred for a complete bibliography of work up to that date).

Because of the exclusive presence of H-Y antigen in the mammalian male (Wachtel *et al.* 1975) and its detection early in embryonic development (Krco and Goldberg 1976) it has become possible to predict the phenotypic sex of the offspring on the basis of embryonic H-Y antigen expression. Epstein *et al.* (1980) exposed a large number of eight-cell stage mouse embryos to H-Y antiserum and complement, and observed that approximately half the embryos showed some degree of cell lysis. Furthermore, of the embryos not showing any cell lysis, 92 per cent karyotyped as females. White *et al.* (1982), and Shelton and Goldberg (1984) did similar sets of experiments, but instead of karyotyping the embryos they transferred them to pseudopregnant recipients, and obtained 82 and 86 per cent females, respectively. In another study, White *et al.* (1983) used monoclonal antibodies to detect H-Y antigen on morula and blastocyst stage mouse embryos using a cytotoxicity assay and an immunofluorescent assay. Of those embryos classified as non-affected by the cytotoxicity test 81 per cent were females, and of those embryos classified as non-fluorescent 83 per cent were female. White *et al.* (1984), using polyclonal antisera raised in mice, recently reported the successful detection of a male-specific factor on preimplantation bovine embryos by the use of an indirect immunofluorescent assay. One hundred and fifty-five embryos were evaluated and 81 (52 per cent) displayed cell-specific fluorescence. When both fluorescent and non-fluorescent embryos were karyotyped, 85 per cent of those fluorescing were karyotyped as males, while 97 per cent of those non-fluorescing were karyotyped as females. Similarly, Brunner *et al.* (1983) and Wachtel (1984), using anti-H-Y monoclonal antibodies in an indirect immunofluorescent assay, obtained results comparable to those of White *et al.* (1984).

It appears that H-Y antigen detection on pre-implantation embryos can be successfully used as a method for selecting the sex of embryos to be transferred to recipients. Even with the use of monoclonal antibodies, however, the accuracy of prediction of the right sex has not gone beyond 87% (Wachtel 1984) and is not much different from the accuracy obtained with polyclonal antiserum (White *et al.* 1984). This seems to indicate that either the monoclonal antibodies used were not very specific and were cross-reacting with other cell surface antigens, or that expression of H-Y antigen on pre-implantation embryos is not an 'all or none' event, but is under a more complex control of expression.

In order to optimize this method of sexing embryos, at our laboratory we tried to prepare high-specific monoclonal antibodies against the H-Y antigen according to techniques initially developed by Köhler and Milstein (1975).

Myeloma cells were fused with spleen cells from female C57BL/6 mice immunized with cells from males of the same highly inbred strain. Positive anti-H-Y antigen clones were detected by enzyme immunoassays based on supernatant fluids of either Daudi-cell cultures or testicular cell preparations taken from normal and sterilized mice, rabbits, and calves, respectively. These fluids served as presumptive sources of H-Y antigen (Nagai *et al.* 1979; Müller *et al.* 1978). Male specificity was ascertained by the fact that the antibodies could be absorbed with spleen cells from male, but not from female mice. Binding of the antibodies to H-Y antigen on the cell surface was confirmed by indirect immunofluorescent assays using both male and female cells, which were obtained from a number of tissues and species. Several cell lines appeared to be positive in all different types of assays, indicating that monoclonal antibodies against mouse H-Y antigen may be useful to identify male and female embryos of, for example, the bovine species.

Based on the binding of the antibodies in the screening assays described, several highly specific monoclonal antibodies have been selected to be used in a fluorescent system. In brief, donor cattle are superovulated and the embryos obtained by flushing of the uterus. The embryos are washed and then exposed to the anti-H-Y antibody conjugated to fluorescein isothiocyanate (FITC). After incubation, the embryos are washed again and evaluated for bound label under the fluorescent microscope. After assignment of H-Y phenotype, embryos are karyotyped in order to confirm correct identification. Preliminary results indicate that a fluorescent assay based on the use of highly specific monoclonal antibodies, will prove useful in identification of the sex of embryos.

Concluding remarks

There is no doubt that an ideal method of control of sex ratio such as separation of X- and Y-bearing spermatozoa, would find widespread acceptance and use. Until recently, evaluation of new methods for separating sperm was hampered by the lack of laboratory tests to check the degree of separation. Nowadays, flow cytometry offers the possibility of evaluating rapidly and on a sound statistical basis the percentages of X and Y sperm in populations supposedly enriched by a variety of procedures. The availability of this technique will certainly speed up research in this field.

At the present time, sexing sperm does not seem to be an immediate prospect. Physical techniques did not have any effect on the sex ratio yet and haploid expression of X- or Y-genes is unlikely to happen. As a future possibility hybridization probes for DNA sequences specific to the X or Y chromosome may have application to develop a reliable method for producing populations of sperm greatly enriched in X- or Y-bearing sperm within the next decade (Epplen *et al.* 1981; Jones and Singh 1981).

In order to sex embryos two actual diagnostic procedures need to be considered, sex chromosomal analysis and detection of the H-Y antigen present on male cells. It is generally agreed that the difficult and time consuming techniques, the requirement for trained cytogeneticists, and moreover the 30–50 per cent survival of sexed embryos are discouraging for commercial embryo sexing by karyotyping. On the other hand, detection of H-Y antigen on male embryos may prove useful in the control of sex ratios in domestic species. Compared with cytogenetic analysis, detection of H-Y antigen with a fluorescent system has the advantage that embryos can be sexed at days 6–12 of gestation, corresponding to the stages at which embryo transfer is practicable. The technique sets no limit on the number of embryos to be sexed, takes only 2 h, can be easily applied by people with little or no experience in immunological techniques, and data suggest that the viability of the embryo is not affected by the manipulations (K. L. White, 1985, personal communication). It is likely that this technique will become widely available within a year or two.

It remains to develop a method that eliminates the need for a fluorescent microscope and that can be used more easily on the farm. In this respect Wachtel (1984) reported the development of an enzyme immunoassay. The principle of such an assay is based on the reaction of enzyme and its substrate to generate a colored product. Anti-H-Y antibodies carrying an enzyme bind with the embryo. Adding the substrate yields a colour demonstrating the presence of H-Y antigen. Enzyme systems allow for simple visual scoring and can be made available in a kit form suitable for on-farm use. In all likelihood it will soon be possible to routinely and effectively select the sex of the offspring in the domestic animal species.

References

Amann, R. P. and Seidel, G. E., Jr (1982). *Prospects for sexing mammalian sperm.* Colorado Associated University Press, Boulder, Colorado.

Betteridge, K. J., Hare, W. C. D., and Singh, E. L. (1981). Approaches to sex selection in farm animals. In *New technologies in animal breeding.* (eds B. G. Brackett, G. E. Seidel, Jr, and S. M. Seidel) pp. 109–25. Academic Press, New York, London.

Billingham, R. E. and Silvers, W. K. (1960). Studies on tolerance of the Y-chromosome antigen in mice. *J. Immunol.* **85**, 14–26.

Brunner, M., Moreira-Filho, C., Selden, J. R., Koo, G. C., and Wachtel, S. S. 1983. H-Y antigen in the early embryo. In *Proc. 2nd symp. adv. topics animal reproduction.* (eds. L. E. L. Pinheiro and P. K. Basrur) pp. 97–111. FCAV-UNESP Grafica, Jabotical SP Brasil.

Bryant, B. J. (1980). Method and material for increasing the percentage of mammalian offspring of either sex. *US patent* 4, **191**, 749.

Eichwald, E. J. and Silmser, C. R. (1955). Transplant. Bull. **2**, 148–9.

Epstein, C. J., Smith, S., and Travis, B. (1980). Expression of H-Y antigen on pre-implantation mouse embryos. *Tissue Antigens* **15**, 63–7.

Epplen, J. T., Sutou, S., McCarry, J. R., and Ohno, S. (1981). Sex determining genes and gene regulation. In *Proc. ORPRC Symp. Primate Reprod. Biol.*, Vol. 1. *Fetal Endocrinology*, ed. M. J. Nooy and J. A. Resko pp. 239–51. Academic Press, New York.

Ericsson, R. J., Langevin, C. N., and Nishino, M. (1973). Isolation of fractions rich in human Y sperm. *Nature* **246**, 421–4.

Garner, D. L., Gledhill, B. L., Pinkel, D., Lake, S. Stephenson, D., Van Dilla, M. A., and Johnson, L. A. (1983). Quantification of the X- and Y-chromosome bearing spermatozoa of domestic animals by flow cytometry. *Biol. Reprod.* **28**, 312–21.

Gledhill, B. L. and Garner, D. L. (1983). Control of mammalian sex ratio at birth. *Symp. Ann. Meeting Am. Assoc. Advancement of Science, Detroit.*

Goldberg, E. H., Boyse, E. A., Bennett, D., Scheid, M., and Carswell, E. A (1971). Serological demonstration of H-Y (male) antigen on mouse sperm. *Nature*, **232**, 478–80.

Hare, W. C. D., Mitchell, D., Betteridge, K. J., Eaglesome, M. D., and Randall, G. C. B. (1976). Sexing two-week old bovine embryos by chromosomal analysis prior to surgical transfer: Preliminary methods and results. *Theriogenology* **5**, 243–53.

Hoppe, P. C. and Koo, G. C. (1984). Reacting mouse sperm with monoclonal H-Y antibodies does not influence sex ratio of eggs fertilized *in vitro*. *J. Reprod. Immunol.* **6**, 1–9.

Jones. K. W. and Singh, L. (1981). Conserved repeated DNA sequences in vertebrate sex chromosomes. *Human Genet.* **58**, 46–53.

Kiddy, C. A. and Hafs, H. D. (1971). Sex ratio at birth prospects for control. Champaign, Illinois, American Society of Animal Science.

King, W. A. (1984). Sexing embryos by cytological methods. *Theriogenology* **21**, 7–17.

Köhler, G. and Milstein, C. (1975). Continuous culture of fused cells secreting antibody of predetermined specificity. *Nature* **256**, 495–7.

Koo, G. C., Stackpole, C. W., Boyse, E. A., Hammerling, U., and Lardis, M. (1973). Topographical location of H-Y antigen on mouse spermatozoa by immunoelectronmicroscopy. *Proc. Nat. Acad. USA* **70**, 1502–5.

Krco, C. J. and Goldberg, E. H. (1976). Detection of H-Y (male) antigen on 8-cell mouse embryos. *Science* **193** 1134–5.

Moustafa, L. A., Hahn, J., and Roselius, R. (1978). Versuche zur Geschlechtsbestimmung an Tag 6 and 7 alten Rinderembryonen. *Tierärztl. Wschr.* **91**, 236–8.

Muller, U., Aschmoneit, I., Zenzes, M. T., and Wolf, U. 1978. Binding studies of H-Y antigen in rat tissues. Indications for a gonad-specific receptor. *Human Genet.* **43**, 151–7.

Nagai, Y., Ciccarese, S., and Ohno, S. 1979. The identification of human H-Y antigen and testicular transformation induced by its interaction with the receptor site of bovine fetal ovarian cells. *Differentiation* **13**, 155–64.

Ohno, S. (1979). *Major Sex-determining Genes.* Springer-Verlag, Berlin.

—— (1982). Expression of X- and Y-linked genes during mammalian spermatogenesis. In *Prospects for sexing mammalian sperm* (eds R. P. Amann and G. E.

Seidel, Jr) pp. 109–13. Colorado Associated University Press, Boulder, Colorado.

Pearson, P.L., Bobrow, M., Vosa, C.G., and Barlow, P.W. (1971). Quinacrine fluorescence in mammalian chromosomes. *Nature* **231**, 326–9.

Pinkel, D., Garner, D.L., Gledhill, B.L., Lake, S., Stephenson, D., and Johnson, L.A. (1983). The proportions of X- and Y-chromosome-bearing spermatozoa in samples of purportedly enriched bovine semen. *J. Anim. Sci.* **57** (Suppl. 1), 366.

—— Lake, S., Gledhill, B.L., Van Dilla, M.A. Stephenson, D., and Watchmaker, G. (1982). High resolution DNA measurements of mammalian sperm. *Cytometry,* **3**, 1–9.

Shelton, J.A. and Goldberg, E.H. (1984). Expression of H-Y antigen on preimplantation stage male mouse ambryos. *J. Reprod. Immunol.* Suppl. S4.

Schneider, U. and Hahn, J. (1979). Bovine embryo transfer in Germany. *Theriogenology* **11**, 63–80.

Wachtel, S.S. (1983). *H-Y antigen and the biology of sex determination.* Grune and Stratton, New York.

—— (1984). H-Y antigen in the study of sex determination and control of sex ratio. *Theriogenology* **21**, 18–28.

—— Koo, G.C., and Boyse, E.A. (1975). Evolutionary conservation of H-Y (male) antigen. *Nature* **254**, 270–2.

—— and Ohno, S. (1980). The immunogenetics of sexual development. *Ann. Rev. Med. Genet.* **3**, 109–42.

White, K.L., Bradbury, M.W., Anderson, G.B., and Bondurant, R.H. (1984). Immunofluorescent detection of a male-specific factor on preimplantation bovine embryos. *Theriogenology* **21**, 275.

—— Lindner, G.M., Anderson, G.B., and Bondurant, R.H. (1982). Survival after transfer of 'sexed' mouse embryos exposed to H-Y antisera. *Theriogenology* **18**, 655–62.

——, ——, —— and —— (1933). Cytolytic and fluorescent detection of H-Y antigen on preimplantation mouse embryos. *Theriogenology* **19**, 701–5.

Zavos, P.M. (1983). Preconception sex determination via intra-vaginal administration of H-Y antisera in rabbits. *Theriogenology* **20**, 235–40.

Zech, L. (1969). Investigation of metaphase chromosomes with DNA binding fluorochromes. *Exp. Cell Res.* **58**, 463.

Discussion

Robertson asked how many early embryos have been sexed by the monoclonal antibody technique. The answer was some 40 embryos. Di Berardino wondered if fluorescent dyes and UV light may damage the embryos. Booman thought not, for several groups have reported similar implantation and survival rates in treated and control groups of mice and sheep. Di Berardino also was concerned that though the embryos may be phenotypically normal they perhaps may carry mutations. No data on this were available on an adequate scale. However, the level of UV light is low and the fluorescing antibody is degraded rapidly in the growing embryo. Moreover, Lovell-Badge pointed out that new image intensifiers allow the use of even lower levels of UV and fluorescence. Booman suggested that future techniques may be enzyme linked and be measured by a simple colour change in the medium. The current embryo sexing has been tested successfully on horse, goat and even fish, and so may apply to many species.

Brem asked when H-Y appears in developing embryos. It is certainly present at the eight-cell stage in the mouse, and is probably detectable at the four-cell stage. It may be expressed earlier, but is hard to detect. Hill wondered why polyclonal antibodies are still being used. Booman replied that monoclonal antibodies are only used in groups where the technology is available. Eventually, all groups will use monoclonal antibodies because of their advantages, especially standardizing over years. The figures presented were for polyclonals since too few determinations have been done by the monoclonal method to get accurate figures.

3
Application models of reproductive manipulation

G. Brem

Abstract

*In our experiments, embryomicrosurgery (EM) was used to produce mono-
zygotic twins and chimerae in cattle. Application models for such twin production include research programmes. For research, a monozygotic – dizygotic
model is proposed. Later, for estimating maternal effects and genetic or
environmental change, splitting allows better methods. In AI and nucleus
breeding programmes genetic improvement will be increased and new applications in breeding test bulls are possible. The genetic superiority of progenies in
herd breeding programmes using embryo transfer (ET)-EM can be doubled.
Chimerae will be used in research to study lethal disorders or for production of
transgenic cattle. Splitting embryos will increase the variety and impact of all
application possibilities of conventional (ET) programmes especially because,
by the use of this technique, a decrease in costs can be expected.*

Introduction

Egg transfer techniques have been commercially exploited in cattle to a
considerable extent in recent years. In addition, since 1981 the micromanipulation technique of cleaving bovine embryos to produce monozygotic twins
has been established (Willadsen 1981; Williams *et al.* 1982; Ozil 1983; Brem *et
al.* 1983). Furthermore, deep freezing of bovine half embryos was described
in recent reports (Lehn-Jensen and Willadsen 1982; Heyman and Chesne
1984).

In our experiments, morulae and blastocysts were split using a mobile
micromanipulation unit. All proceedings, collection, splitting, and transfer
of embryos were done on the farm under practice conditions. In this way we
produced 35 pairs of twins in total. In our routine programmes embryo survival rate was about 105 per cent and twinning rate 33 per cent.

The objectives of this paper are to present some application models for
embryo splitting in research, AI and herd programmes. In addition, some
future aspects will be considered which would result from the routine
production of bovine chimerae.

Monozygotic (MZ) twins arisen from a natural phenomenon have been of great importance in animal breeding investigations in the past. Production of MZ twins using embryo-microsurgery renders additional applications in research and breeding programmes.

Animal breeding research programme

Studies on monozygotic cattle twins have a long tradition, because of the greater efficiency of monozygotic twins in comparison to ordinary animals. Twins collected from the population are not available until an age of several weeks when the positive diagnosis of monozygosity is made. In contrast calves derived from split embryos can be used for examinations immediately after birth or even before.

Advantages of the production of MZ using embryo-microsurgery (EM) are:

— parents can be chosen in advance;
— no extensive diagnosis of monozygosity is necessary;
— environmental effects (intrauterine, maternal, perinatal) can be modified for each part of a pair by separating them before transfer or immediately after parturition or by freezing one demi-embryo;
— number of pairs and time of birth can be determined according to the requirements of the programme and will to a much lower extent depend on accidental events;
— several mono- and/or dizygotic pairs, full sibs, and maternal or paternal half sibs of the same age can be produced;
— pairs of twins are available for investigation during the pregnancy period;
— the exact number of pairs needed can be produced much easier than obtained by collection

Disadvantages are variations in the results of the embryo-microsurgery procedure in certain specific cases. Success rates can only be predicted if a large number of embryos is manipulated. Long-term efforts are necessary in order to establish a trained team and set up all the equipment necessary for doing highly successful embryo splitting and embryo transfer (ET).

The combined transfer of the two demi-embryos of a monozygotic pair to two recipients, each together with one demi-embryo of another monozygotic pair to the contralateral uterine horn results in a highly effective mono-zygotic–dizygotic model for investigations (Fig. 3.1). In our laboratory we started some projects regarding physiological, clinical and immunological parameters of newborn twin calves before the intake of colostrum. Mitogen induced stimulation of lymphocytes in newborn monozygotic and dizygotic twin calves was measured. With only few calves produced by splitting embryos it could be shown, that genetic influences are rather weak as compared to environment influences (Buschmann, *et al.* 1985).

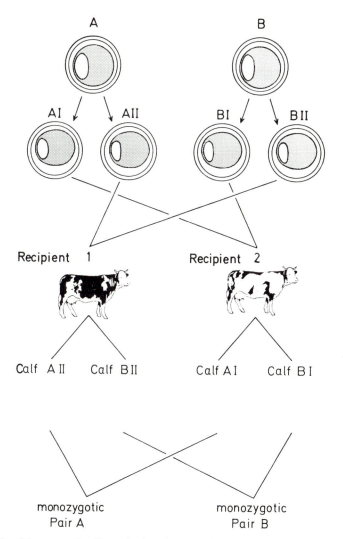

Fig. 3.1. Monozygotic–dizygotic situation established by cross transfer of two split embryos into two recipients.

Some other possible areas for studies with monozygotic twins are genotype-environment interactions, liability, maternal effects, and the measurement of genetic and environmental trends.

Maternal effects

Problems in estimating maternal effects have been discussed by Willham (1979). The phenotypic variance for a trait is composed of the sum of direct and maternal, additive and dominant, mitochondrial and environmental components. Maternal effects can be divided in:

1. Maternal genetic effects
2. Extrachromosomal effects
3. Effects of the uterine environment
4. Post-partal maternal environment

The composition of covariances between conventional relatives and relatives produced by ET and embryomanipulation is shown in Table 3.1 under consideration of these four effects. Some components can be estimated as differences between selected covariances. All components will be estimated simultaneously by least square procedures applied to the groups of animals in Table 3.1.

Table 3.1.   Components of covariances between relatives from ET-EM

Covariance	Variance Component							
	$V_{Ga(d)}$	$V_{Ga(m)}$	$V_{Gd(d)}$	$V_{Gd(m)}$	$V_{Mi(m)}$	$V_{Ut(m)}$	$V_{E(m)}$	$V_{E(d)}$
$COV(MZ_{1R, 1E})$	1	1	1	1	1	1	1	1
$COV(MZ_{1R, 2E})$	1	1	1	1	1	1	0	0
$COV(MZ_{2R, 2E})$	1	1	1	1	1	0	0	0
$COV(MZ_{GT, 2R})$	1	1	1	1	0	0	0	0
$COV(MZ_{GT, 1R})$	1	1	1	1	0	1	1	1
$COV(FS_{MZ:DZ}, 1E)$	1/2	1	1/4	1	1	1	1	1
$COV(FS)$	1/2	1	1/4	1	1	1	1	0
$COV(FS,_{ET})$	1/2	1	1/4	1	1	0	0	0
$COV(M, PR)$	1/2	1/2	0	0	1	0	0	0
$COV(mHS)$	1/4	1	0	1	1	1	1	0
$COV(F, PR)$	1/2	0	0	0	0	0	0	0
$COV(pHS,_{ET, 1R})$	1/4	0	0	0	0	1	1	1
$COV(pHS)$	1/4	0	0	0	0	0	0	0
$COV(NR_{ET, 1R})$	0	0	0	0	0	1	1	1
$COV(NR)$	0	0	0	0	0	0	0	1

Ga = additive genetic effects; Gd = dominant genetic effects; Mi = mitochondrial effects; Ut = uterine effects; E = envitronmental effects; (d) = direct effects; (m) = maternal effects; MZ = monozygotic twins; DZ = dizygotic twins; R = recipient (1R = both demi-embryos transferred into one recipient); E = environment (1E = common environment for all calves); GT = genome transfer; FS = full sibs; HS = half sibs; M = mother; F = father; PR = progenies; NR = non-relatives.

$V_p = V_{Ga(d)} + V_{Ga(m)} + V_{Gd(d)} + V_{Gd(m)} + V_{Mi(m)} + V_{Ut(m)} + V_{U(m)} + V_{U(d)}$

This composition of phenotypic variance does not include interactions or epistatic effects.

Estimation of genetic and environmental change

Methods of estimating genetic change are comparison of alternatives selection schemes, constant environment, replication of the same genetic material in successive generations or analysis of field records (Hill 1972). The use of frozen semen and frozen embryos in measuring genetic trends in performance has been proposed by Smith (1977).

The variance (V) of the mean in setting up a pool of split embryos will be:

$$V = V_p \{0.5 + 0.5\,(h^2 + c^2 + e^2) + (z-1)(0.5h^2 + c^2) + (m-1)\,z\,0.25h^2\}/zvm$$

where z is the number of pairs per dam, m is the number of dams per sire, v is the number of unrelated sires, and V_p is the phenotypic variance.

The production of monozygotic twins of different age gives a better means for estimating genetic and environmental changes. One demi-embryo will be transferred immediately after splitting the other demi-embryo will be stored in liquid nitrogen. Some generations later this demi-embryo will be thawed and transferred too. The situation after six generations is shown in Fig. 3.2. The production of MZ of different age is shown in Fig. 3.4.

Comparing the performance of both individuals of one monozygotic pair in different generations the estimate for the environmental change is

$$\Delta U = (MZ_t - MZ_0)/t$$

and the genetic change is

$$\Delta G = (BP_t - t\Delta U)/t,$$

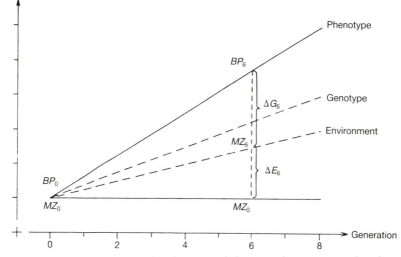

Fig. 3.2. Measuring genetic and environmental changes using monozygotic twins of different age.

where MZ_0 is the performance of the 1st member of a pair in generation O, MZ_t is the performance of the 2nd member of a pair in generation t, and BP_t is the increase in the average of performance of the breeding population after t generations.

The increase of environmental improvements and genetic gain will not be constant as shown in Fig. 3.2. Thus, every one or two generations new estimations will be helpful to find out these differences. Comparing traits not under selection will show their correlated changes.

AI breeding programmes

Nicholas and Smith (1983) proposed the use of embryo transfer and embryo splitting in dairy cattle involving females after one lactation (adult scheme) or 1-year-old females (juvenile scheme). They found a genetic improvement some 30% above that by a conventional AI breeding programme.

Assuming a selection intensity in the path dam – son between 10 and 1 per cent, in a German AI breeding programme we estimated the annual genetic gain on this path. Figure 3.3 shows the progress comparing a programme only using ET for different numbers of progenies per dam. The base plan is the conventional AI programme with selection on a single trait.

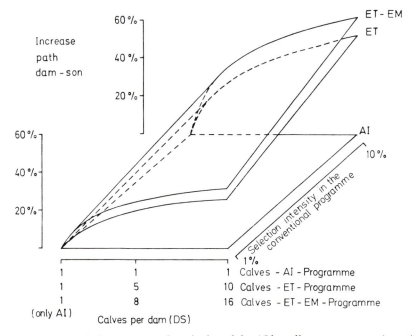

Fig. 3.3.   Genetic improvement in a single trait by AI breeding programmes (genetic gain in standard deviations).

Similar to the results of Nicholas and Smith (1983), the effect of splitting on genetic change is quite small. On the other hand, the economic value of the additional calves produced by microsurgery can help to reduce costs and in this way to establish such programmes.

Embryo splitting provides new applications in breeding test bulls by improving test possibilities and reducing the number of waiting bulls. In comparison to testing an individual, testing a monozygotic pair raises the accuracy of evaluation by 35–40 per cent if heritability is below 0.10, and 20–29 per cent if heritability is between 0.20 to 0.40. According to Van Vleck (1981) the accuracy of evaluation of single records on two clones becomes:

$$r_{AI} = \{2h^2/(1 + h^2)\}^{1/2}$$

Another use of monozygotic twins in cattle breeding is the evaluation of characteristics which cannot be measured in the breeding animal itself, for example, the carcass value of young bulls. Exact evaluation of carcass value and quality requires slaughtering of the animal.

Another example is the resistance to diseases. Evaluation of specific or general disease resistance requires the iatrogenic infection of the animal or the application of immunological tests. Both treatments are often incompatible with later usage of the bulls for breeding (Kräußlich and Brem 1983). If monozygotic pairs are available both could be tested in fattening traits, and in addition one yields information on carcass quality and performance or resistance to disease whereas the other can be used in breeding.

Freezing of demi-embryos can be used to produce monozygotic pairs of different age (Fig. 3.4). After testing the first calf the second demi-embryo will be thawed and transferred. If this transfer results in a pregnancy, bull I could be slaughtered. By the time the performances of the daughters of bull I are available, bull II will be in best breeding condition and can be used for insemination if the breeding value of the monozygotic pair is good enough.

Herd breeding programmes

The cow dam selection is carried out within herds by individual breeders. In a self-recruiting herd ET allows replacement of progenies of the poorest cows by the progenies of the best cows, i.e. the best cows must be selected as donor cows and the others will serve as recipient cows. The added offspring superiority within a herd of 100 cows, (six eggs recovered per cow, 17 donor cows) would be 0.31 $r_{IA}s_A$-units (Petersen 1978).

The genetic superiority of the female progenies (G_{FP}) depends on the superiority of the donors (G_{Don}), the number of calves produced by egg-transfer and embryo-splitting (n_{CET}), and the number of cows in a herd (n_H).

$$\Delta G_{FP} = \Delta G_{Don} \cdot 0.5 \, n_{CET}/n_H$$

The genetic gain depends directly on the product of pregnancies per embryo

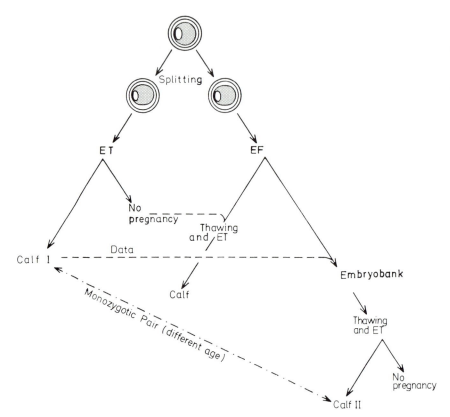

Fig. 3.4.   Freezing of demi-embryos to produce monozygotic pairs of different age.

and pairs of twins per embryo. In Fig. 3.5 genetic superiority of calves derived from conventional ET programmes are combined with different success rates of splitting embryos. In many cases the genetic gain can be doubled.

Nucleus breeding programmes

Recently, Hill (1971) has demonstrated that a herd nucleus may supply superior male genotypes and Hinks (1977) developed a nucleus herd selection programme in dairy cattle breeding. Embryo transfer and embryo splitting can be used in a nucleus programme in two different ways.

1. During the foundation of the nucleus, in order to increase the superiority of the selected genotypes.
2. During the subsequent selection, in order to increase selection intensities and accuracies.

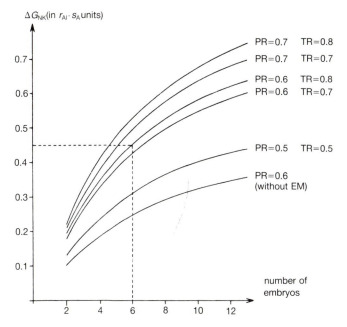

Fig. 3.5.   Genetic superiority of progenies in herd programmes using ET–EM. (PR = pregnancy rate, TR = twinning rate, NK = offspring)

In German dairy and dual purpose cattle programmes the nucleus system will be of limited success for raising the performances in milk and meat, because the current standard in conventional programmes in these traits is relatively high. On the other hand, progress may be accelerated if new traits like fertility, feed conversion, or resistance against diseases will be included in the breeding programmes, because measuring these characters will be easier in the nucleus than in the population. Furthermore, nucleus breeding programmes could be a good strategy for genetic improvement in 'third world' countries which lack the money, expertise, and structure required for operating an effective improvement programme with modern AI and milk recording services in the whole population.

Chimera

Recently, experimental chimerism in sheep (Fehilly *et al.* 1984a) and the production of interspecific chimerae between sheep and goat (Fehilly *et al.* 1984b) has been reported. So far, cattle chimerae have been produced by aggregating blastomeres, derived from surgically flushed eight-cell stage embryos, which then were cultured in ligated sheep oviducts, before their transfer to final recipients (Fehilly *et al.* 1984a). The blastocyst injection technique was used in cattle. None of the resulting calves were overtly chimeric, but there was evidence of chimerism of internal tissues in one animal (Summers *et al.* 1983).

We have produced cattle chimerae from non-surgically recovered morulae of day 5 to 6.5 by combining four halves of two parent embryos of different breeds and transferring the aggregated embryos non-surgically to synchronized recipients (Brem *et al.* 1984a; Brem 1985, unpublished data).

Chimerae have been used for investigation on growth regulation and effects of heterosis in mice (Falconer *et al.* 1981). Figure 3.6 shows a suitable, but expensive design for simultaneous evaluation of cellular, uterine and external environmental effects in cattle. If genome transfer between the pure breeds were done, cytoplasmatic effects could also be tested.

Another application model using chimerism could be the study of lethal disorders similar to projects in mice (McLaren 1976). Chimeric embryos out of defective and normal embryos may give rise to surviving individuals. Chimeric calves produced by this method could help to find the genetic and developmental causes of lethal disorders. An actual problem to be investigated by this procedure would be the lethal defect Arachnomelie in the Mid-European Brown-Swiss populations. This syndrome has a gene frequency of almost 5% and it should be helpful to obtain living calves with the defect in order to figure out a testing system for diagnosis of heterozygotes, because about 10% of the population carries the recessive gene (Brem *et al*, 1984b).

A future project with chimerism in cattle could be the production of transgenics (Fig. 3.7). All the techniques necessary have been successfully developed in mice (Mintz 1979), but it will take a long time to get ec-cells out of cattle embryos. The genome of these cells should be derived from matings of sires to dams intended for breeding sires because this genome would get widely spread in the population, assuming that the whole procedure was successful. The male ec-cells would be transformed with new genes and after testing integration and expression of these genes in the cells used for the production of chimerae. If the transgenic cells were incorporated into the gonads of the chimerae, transgenic paternal half sibs could be produced by mating the chimeric bull to the population.

References

Brem, G., Kräußlich, H., Szilvassy, B., Kruff, B., Lampeter, W.W., and Graf, F. (1983). Zur Erzeugung eineiiger Rinderzwillinge durch Embryo-Mikrochirurgie. *Berl. Münch. tierärztl. Wschr.* **96**, 153–7.
——, Tenhumberg, H., and Kräußlich, H. (1984a). Chimerism in cattle through microsurgical aggregation of morulae. *Theriogenology* **5**, 609–13.

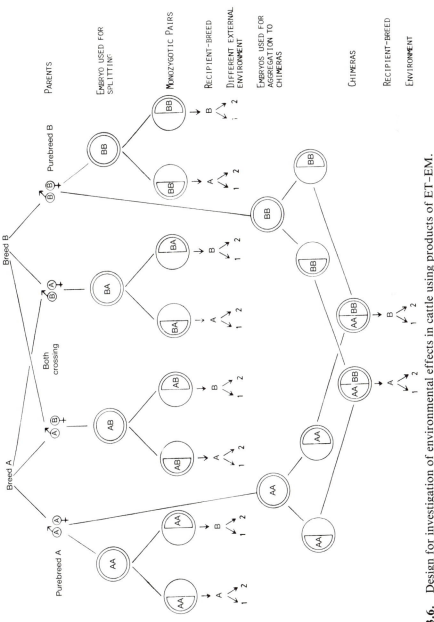

Fig. 3.6. Design for investigation of environmental effects in cattle using products of ET–EM.

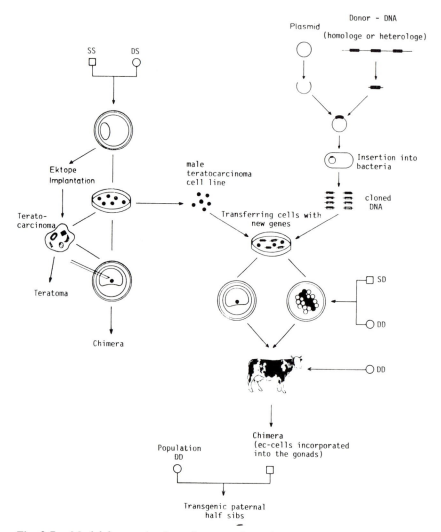

Fig. 3.7. Model for production of transgenic cattle with chimerae of embryo and teratocarcinoma cells.

——, Wanke, R., Hondele, J., and Dahme, E. (1984b). Zum Auftreten der Arachnomelie in der Braunvieh x Brown-Swiss Population Bayerns. *Berl. Münch. tierärztl. Wschr.* **97**, 393–7.

Buschmann, G., Brem, G., Kleinschmidt, A., and Meyer, J. (1985). Mitogen stimulation of lymphocytes from bovine twins reared in different maternal environment. *Zbl. Vet. Med., Reihe B.* (in press).

Falconer, D. S., Gauld, I. K., Roberts, R. C., and Williams, D. A. (1981). The control of body size in mouse chimeras. *Genet. Res. Camb.* **38**, 25–46.

Fehilly, C. B., Willadsen, S. M., and Tucker, E. M. (1984a). Experimental chimerism in sheep. *J. Reprod. Fertil.* **70**, 347–51.

——, ——, and —— (1984b). Interspecific chimerism between sheep and goat. *Nature* **307**, 634–6.

Heyman, Y. and Chesne, P. (1984). Freezing bovine embryos: survival after cervical transfer of one half, one or two blastocysts frozen in straws. *Theriogenology* **21**, 240.

Hill, W. G. (1971). Investment appraisal for national breeding programmes. *Anim. Prod.* **13**, 37–50.

—— (1972). Estimation of genetic change. I. General theory and design of control populations. *Anim. Breed. Abstr.* **40**, 1–12.

Hinks, C. J. M. (1977). The development of nucleus herd selection programmes in dairy cattle breeding. *Z. Tierz. ZüchtBiol.* **94**, 44–54.

Kräußlich, H. and Brem, G. (1983). Future aspects of micromanipulation with embryos for cattle breeding. *S.A. J. Anim. Sci.* **13**, 286–91.

Lehn-Jensen, H. and Willadsen, S. M. (1982). Deep-freezing of cow 'half' and 'quarter' embryos. *Theriogenology* **19**, 49–54.

McLaren, A. (1976). Mammalian Chimeras. In *Developmental and cell biology series, No. 4* (eds M. Abercrombie, R. Newth, and J. G. Torrey). Cambridge University Press.

Mintz, B. (1979). Teratocarcinoma cells as vehicles for introducing mutant genes into mice. *Differentiation* **13**, 25–27.

Nicholas, F. W. and Smith, C. (1983). Increased rates of genetic change in dairy cattle by embryo transfer and splitting. *Anim. Prod.* **36**, 341–53.

Ozil, J. P. (1983). Production of identical twins by bisection of blastocysts in the cow. *J. Reprod. Fertil.* **69**, 463–8.

Petersen, P. H. (1978). Breeding aspects of egg transplantations exploited in purebred dairy and dual-purpose cattle. *29th Ann. Meeting Eur. Ass. Anim. Prod. Stockholm*, 5–7 June 1978.

Smith, C. (1977). Use of stored frozen semen and embryos to measure genetic trends in farm livestock. *Z. Tierz. ZüchtBiol.* **94**, 119–27.

Summers, P. M., Shelton, J. N., and Bell, K. (1983). Synthesis of primary bos taurus – bos indicus chimeric calves. *Anim. Prod. Sci.* **6**, 91–102.

Van Vleck, L. D. (1981). Potential genetic impact of artificial insemination, sex selection, embryo transfer, cloning and selfing in dairy cattle. In *New technologies in animal breeding* (ed. B. G. Brackett, C. E. Seidel jun., and S. M. Seidel), pp. 221–42. Academic Press, London.

Willadsen, S. M. (1981). Micromanipulation of embryos of the large domestic species. In *Mammalian Egg Transfer*, pp. 185–210. CRC Press, Florida.

Willham, R. L. (1979). Problems in estimating maternal effects. *30th Ann. Meeting Eur. Ass. Anim. Prod. Harrogate*, 23–26 July, 1979.

Williams, T. J., Elsden, R. P., and Seidel, G. E. (1982). Identical twin bovine pregnancies derived from bisected embryos. *Theriogenology,* **17**, 114.

Discussion

The value of chimeras was questioned because of the problem of random integration. Were there rules for getting cells from one parent into the germ line? The answer seemed to be that it would be extremely nice to know what governed the outcome and until then one had to wait until the right combination was obtained by chance. It was pointed out that the cross-nursing system proposed would be handicapped by problems with free martins, and this was agreed to be so until such time as embryos could be sexed. (Sexing would also be useful in producing chimeras.)

The scheme for using split embryos in the testing of AI bulls caused debate about the risk of non-conception of the second frozen embryo. This was agreed and the key to the system was to wait until pregnancy had been established before slaughtering the first twin. The risks from then on were considered lower and more tolerable. On challenge, it did seem that the collection of more semen and freezing it, might be a cheaper alternative.

The use of identical twins to monitor genetic change was discussed and the need to calculate the appropriate number of twin pairs for measurement was emphasized. Since only the genetic element was removed by the use of such individuals, and this for many characters was only 25% of the total, then surprisingly high numbers were required merely to overcome measurement errors. Hill suggested that the main use for identical twins would be non-genetic experiments and this was agreed with the emphasis being laid on the ability to select parents for particular characteristics to make them appropriate to the experiment in hand.

The success rates with fresh and frozen embryos, both with and without splitting were discussed. Brem had not observed the decline in viability observed after 4 h in Denmark, but had observed that splitting of eggs should take place on the farm where the recipients were located since carrying out this operation in a laboratory and then transporting the split eggs had given poor results. The importance of the culture medium was also indicated and the problems of freezing eggs after splitting them was discussed. There was discussion on the costs of embryo transfer resulting both from lower conception rates, giving a longer dry period, and, in dairy animals, a decline in milk production due to hormone treatment.

4

Embryo transfer in the genetic improvement of dairy cattle

L. G. Christensen and T. Liboriussen

Abstract

The effect of using multiple ovulation and embryo transfer (MOET) in current cattle breeding schemes has rather small effect on the rate of genetic progress. Utilization of MOET in future cattle breeding schemes should aim at producing genetically outstanding bull dams and sires. Establishment of special nucleus herds with use of MOET, short generation interval, performance test of females, and index selection could be important tools to increase the genetic progress from selection of bull dams. If progeny testing of bulls takes place, open nucleus herds with use of MOET and bull sire inseminations would be beneficial. Use of physiological predictors in future schemes is discussed. A project, which incorporates MOET in a breeding scheme for dairy breeds, is briefly described.

Current schemes

Embryo transfer (ET) can contribute to the rate of genetic improvement, mainly by increasing the possible selection intensity in the two paths: cows to produce replacement cows (CC), and cows to produce breeding bulls (CB).

High costs of ET implies that ET is unlikely to become of practical importance for production of replacements. The contribution to the genetic progress from ET in this path will consequently be negligible, unless the costs of ET are reduced by a large multiple (Van Vleck 1981). This is unlikely to occur, considering the complexity of the operation.

In relation to breeding, ET or multiple ovulation and ET (MOET) has consequently mainly been considered as a tool for production of breeding bulls. Theoretical considerations indicate that it might be possible to increase the rate of genetic progress in a dairy population by 10–15 per cent in a conventional breeding scheme, by practising MOET on the very best cows, (McDaniel and Cassell 1981; Künzi 1984).

In practice, however, the genetic difference between sons of extremely high indexed bull dams, and sons of more ordinary bulls dams, has been much less than expected (Trodahl and Syrstad 1977). This indicates, that the breeding values of very high yielding and promising cows are overestimated.

37

Even if we could avoid preferential treatment, as long as breeding values are based on field records, the breeding value of individual cows can only be estimated with moderate accuracy. It can therefore be concluded that ET will only have small effect on the rate of genetic progress in a conventional breeding scheme.

Future schemes

This does not mean, that MOET can't be utilized to increase genetic progress in dairy cattle. What it does imply is that the traditional breeding schemes must be changed in order to make better use of the new technique. In order to obtain this, the following possibilities should be considered:

— preselection of females, based on pedigree indexes;
— establishment of nucleus herds, with performance test of females for dairy traits;
— use of physiological predictors (traits) in the selection of males and females.

The juvenile MOET scheme, proposed by Nicholas (1979) and Nicholas and Smith (1983), is a extreme example of utilization of pedigree information. This scheme involves nucleus herds with selection among transferred males and females at sexual maturity (12–13 months of age) based on first lactation of their dams. The generation interval (L) for both sexes then is less than 2 years ($L = 1.83$).

We use one example from Nicholas and Smith (1983). It was given that $h^2 = 0.25$, $y =$ eight embryos/donor; $x =$ eight mates/sire, $i_{\varphi\varphi} = 1.27$ (25 per cent selection); $i_{\sigma\sigma} = 1.65$ (12.5 per cent selection), and $L_{\varphi\varphi} = L_{\sigma\sigma} = 1.83$.

Selection is based on the first lactation of dams. The accuracy of selection (r^2_{IG}) then becomes:

$$r^2_{IG} = 1/4 \; h^2 = 0.0625.$$

In spite of this low accuracy, the annual genetic progress (ΔG) in phenotypic units is:

$$\begin{aligned}
\Delta G_y &= r_{IG} \, (i_{\varphi\varphi} + i_{\sigma\sigma}) \, (\sigma G/\sigma P)/(L_{\varphi\varphi} + L_{\sigma\sigma}) \\
&= 0.25 \, (1.27 + 1.65) \, 0.5/(1.83 + 1.83) = 0.100 \; \sigma_p
\end{aligned}$$

If C.V. = 20 per cent (as for milk yield) this corresponds to a genetic improvement of 2 per cent per year, which is similar to the expected progress from an optimal traditional breeding scheme.

If a complete pedigree index (including information on the sibs of the ancestors) is used, the genetic progress would be even higher.

In the example above it has been assumed that all information, used for the selection of males as well as females, was obtained in the nucleus herds. We shall now consider some further implications of establishment of nucleus herds, and some modifications that can increase the genetic effect of MOET.

Nucleus herds as a performance test station for cows

In nucleus herds the environment can be controlled, preferential treatment of cows can be avoided, more characteristics can be recorded, and records can be more intensive than in practice. This implies that the heritability of traits will be higher than those observed in the population. Higher heritability of traits are especially important for the accuracy of estimated breeding values of cows, which mainly are based on own performance. The nucleus herds might consequently be considered as performance test stations for cows.

Progeny testing of MOET bulls

In countries with well established test procedures for young bulls, based on progeny testing in the field, very high accuracies are obtained of breeding values of bulls. Under such circumstances, the genetic progress from the path from bulls to bulls (BB) are likely to be higher, when progeny tested bulls are used as bull sires, instead of young bulls, even though the latest are produced by MOET from the best cows in nucleus herds (Christensen 1984).

The difference between the average breeding value of the best proven bulls (PB), and the average breeding value of young MOET bulls (YB) available at the same time can be written as

$$BV_{PB} - BV_{YB} = i\, r_{IG}\, \sigma G - D_L \Delta G_Y$$

where i is the selection intensity of proven bulls, r_{IG} is the correlation between the selection criteria (I) and the true breeding value (G), D_L is the difference between the generation interval of PB and YB, and ΔG_Y is the genetic gain per year in the nucleus herds.

If $h^2 = 0.25$ and C.V. $= 20$ per cent the selection intensity (i) at which the MOET-YB and the MOET-PB available at the same time are of the same genetic level, can be written as:

$$i = 0.1\, D_L \Delta G\% / r_{IG}$$

where $\Delta G\%$ = percent genetic gain per year.

Table 4.1 shows the percentage selection at which the genetic value of the selected MOET-PB tested on 65 daughters ($r_{IG} = 0.9$) are equal to the genetic level of the MOET-YB available at the same time. It is seen that even with extremely high rate of genetic gain in the nucleus herds, 20 per cent selection is enough, unless the generation interval of the proven bulls is very long.

Table 4.1.   Percentage selection among MOET-PB to get equal genetic level of the selected MOET-PB and the MOET-YB used at the same time (h^2 = 0.25, C.V. = 20% and r_{IG} = 0.90)

$\Delta G\%$ in the nucleus	Difference in generation interval between PB and YB		
	4 years	5 years	6 years
1.0	74	66	59
1.5	59	48	37
2.0	44	32	22
2.5	32	20	12
3.0	22	12	6

Progeny testing of bulls in the population also has the advantage that it can include traits with low heritability, and still yield precise estimates of breeding values.

Open nucleus herds

It is obvious, that the nucleus herds should be founded on female progeny of the best proven bulls (BB) and the best cows in the population (CB). MOET of the best cows in the population should be considered, in order to get as high initial genetic level of the nucleus herds, as possible.

If a conventional breeding scheme is operated in the population, parallel with the nucleus breeding scheme, bulls from the nucleus herds should be tested against bulls from the conventional scheme. If inbreeding is negligible, the bulls used in the nucleus herds should always be the best, no matter where they come from. The same principle should apply for donors, although it will be difficult to compare cows, tested in the nucleus herds, with cows from outside.

In Fig. 4.1 various alternatives are compared. It is assumed that the optimal conventional breeding scheme, given by Christensen (1984), is operating in a population of 600,000 cows. The genetic progress ($\Delta G\%$) is measured as per cent of the genetic level of CC in year 0. $\Delta G\%$ per year in the conventional scheme is 2.3.

Curve 1:   Genetic level of CC if the previous breeding system continues ($\Delta G\%$ = 1.0 per year).

Curve 2:   Genetic level of CC in the optimum scheme.

Curve 3:   Genetic level of CB in the optimum scheme.

Curve 4:   Selected females from a juvenile MOET scheme. Eight progeny per donor and eight mates per sire. Generation interval 1.83 years. Index selection. Twenty-five per cent selection of females and 12.5

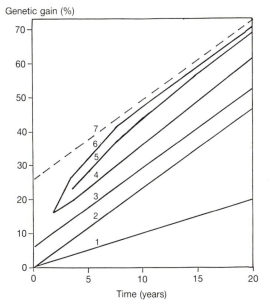

Fig. 4.1. The genetic value of selected females in different breeding schemes (see text).

of males. Genetic superiority (over contemporaries) 4 per cent for females and 5.5 per cent for males. $\Delta G\% = 2.5$ per year.

Curve 5: Genetic level of selected females from an open adult MOET scheme. Generation interval = 3.67 years. Index selection of females based on their own 1st lactation and 1st lactation of their dams. Genetic superiority of selected females 6.8 per cent. Males are BB from the conventional scheme.

Curve 6: Genetic level of selected females from an open juvenile MOET scheme. Generation interval 1.83 years. Selection on the 1st lactation of the dams. Genetic superiority of selected females 3.2 per cent. Males are BB from the conventional scheme.

Curve 7: Genetic level of BB from the conventional scheme.

It is seen that after 10 years the genetic level of the selected females from the open nucleus is 50 per cent higher than that of CC in basic year, and 18 per cent higher than the genetic level of CB available at the same time. At equilibrium the genetic progress in the open nucleus herds is equal to the genetic progress of BB, independent of the female selection.

The genetic progress in the open nucleus can be accelerated further by replacing some of the conventional BB by selected progeny tested MOET bulls. Compared to a MOET scheme in closed nucleus herds, the open nucleus schemes are much less sensitive to inbreeding and random drift.

Physiological predictors

Another possibility of increasing the genetic progress is to utilize physiological predictors measured on the candidates of selection during their first year of life (Walkley and Smith 1980; Smith 1981). The effect of including a physiological index depends on the heritability (h_p^2) of the index, and especially on the genetic correlation (r_G) between the index and the trait, we wish to improve.

This appears clearly from Table 4.2, which is based on the assumption, that the above described juvenile scheme is supplemented with a physiological test on both sexes.

Table 4.2.   The expected rate of genetic gain (ΔG) by including a physiological predictor (I_p) in a juvenile MOET scheme (Nicholas and Smith 1983). It is assumed that I_p is measured on both sexes with heritability (h_p^2), and that the genetic correlation between I_p and the selection criterion is r_G

h_p^2	r_G	Accuracy of the selection index r_{IG}^2	ΔG/year %	Percentage increase of ΔG
0.20	0	0.0625	2.00	0
0.20	0.20	0.0695	2.11	5
0.20	0.50	0.1066	2.61	31
0.20	0.80	0.1760	3.36	68
0.30	0.20	0.0731	2.16	8
0.30	0.50	0.1287	2.87	43
0.30	0.80	0.2333	3.86	93
0.50	0.20	0.0801	2.26	13
0.50	0.50	0.1732	3.33	66
0.50	0.80	0.3495	4.72	136

It is seen that if $h_p^2 = 0.2$, and $r_G = 0.5$, the genetic progress will increase by 31 per cent. If $h_p^2 = 0.3$, and $r_G = 0.8$, the progress will nearly be doubled.

In the adult scheme, proposed by Nicholas and Smith (1983), the effect of including physiological predictors will be less on the female side, but still be large on the male side.

The Danish biological and physiological breeding project

A new breeding project for dairy breeds is initiated this summer in Denmark. It has been planned, and will be guided by the National Institute of Animal Science.

The project can be seen as an attempt to introduce new technologies and methods in dairy cattle breeding such as:

— Multiple ovulation and embryo transfer.
— Nucleus herds with performance testing for dairy characteristics.
— Use of physiological traits for prediction of breeding value for production
  traits.

The project will run parallel to the conventional breeding scheme. The aim is to produce 15–20 per cent of the bulls entering the national A.I. programme, i.e. approximately 100 bull calves per year.

It is a sort of 'joint venture' between The Federation of Danish A.I. Societies, the breed organizations — Danish Friesian, Red Danish, Danish Jersey, and Danish Red and White — and The National Institute of Animal Science, and The Royal Veterinary and Agricultural University in Copenhagen. The A.I. organizations covers all expenses in relation to the procurement of embryos from external donors, and will be the owner of the nucleus herds, and of the bulls produced in the project. An outline of the project is given in Fig. 4.2.

The project starts with a screening of the native cow populations for the best cows. The screening will continue in the years to come. The purpose is to detect highly superior cows, which can serve as donors of embryos. They are called external donors. A contract is made with the owners of these cows, which gives the project the right to superovulate the cows, inseminate them with semen from bulls of the projects choice, and collect the embryos. In the first years an average of five embryos per donor is anticipated. In total approximately 500 embryos are needed per year.

The semen will be from the best bulls within the breeds, or from other breeds which are accepted by the respective breed organizations. The number of bulls used, will be restricted to two or three per breed and year.

The embryos will be transferred to recipients on three experimental stations (the nucleus herds). There will be 240 cows plus approximately 100 heifers, which all can serve as recipients in the early phases of the project. The embryos will, as far as possible, be transferred within 3 h after collection. When this is not possible, the embryos will be frozen and transferred, when it is convenient.

Male calves from the nucleus herds will be performance tested at the performance test stations, together with male calves from the conventional breeding programme. In addition to the ordinary measurements of feed consumption, growth, muscularity (ultrasonic scanning), and registration of diseases, calves from the project will be blood sampled, in order to measure the concentration of various hormones and metabolites, eventually after subjection to physiological challenges. The selection after the performance test will, however, not consider the physiological traits, until we have established reliable and meaningful relationships between the physiological traits and production traits.

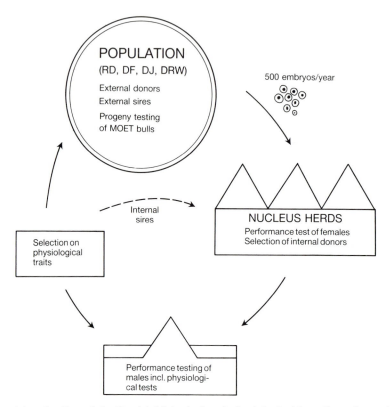

Fig. 4.2. Outline of the Danish biological and physiological breeding scheme (see text).

The best 50–60 per cent of the performance tested bulls will be progeny tested for dairy traits in the ordinary progeny testing programme. The bulls from the project will hence compete with young bulls from the conventional breeding programme. After some years we will be able to estimate the correlation between the physiological traits measured on the young bulls and their breeding value of production and other traits, measured on their daughter groups in the field. These results will serve as guidelines, as to which physiological traits are useful for breeding purposes. If we can improve the accuracy of breeding indexes of the young bulls (by including information about physiological traits), the best young bulls might be better than the selected bulls from the conventional breeding scheme. This is, however, not likely to occur in this decade.

The female calves will remain at the experimental stations, where they will be raised under standardized conditions. Various developmental parameters will be recorded, but no culling will be practised among the heifers. At the age of 18 months they will be used as recipients.

During the first 2 months of their first lactation, they will be subject to intensive examinations, including calving performance, feed intake, milk yield, quality of feet and legs, udder and teats, interval from calving to first oestrus, and diseases. These records on the individual cows will be combined with information on their relatives in order to obtain an early estimate of their breeding value. The high initial breeding values of cows entering the nucleus herds, combined with the relative high accuracy of testing in the herds makes it likely to find young cows in the nucleus herds, which will be better than the best cows in the population. At this stage of the project some of the external donors therefore will be replaced by internal donors (ID). They will be superovulated and serve as donors in competition with the external donors.

In connection with the experiment, a unit with facilities for embryo manipulation will be established. This will facilitate our possibilities to incorporate and utilize new technologies when available. During the whole experimental period we intend to carry through model calculations to determine the optimal number of external and internal donors, embryos per donor, mates per sire, age at selection, selection criterion, etc. The experimental plan will continuously be adjusted according to our findings.

References

Christensen, L. G. (1984). *Breeding problems in small cattle populations*. 35th Meeting EAAP. Den Haag. 18 pp.

Künzi, N. 1984. Die mögliche Nutzung des Embryotransfer in Rinderzuchtprogrammen. *Schr. Schweiz. Verein. Tierz.* **63**, 16–26.

McDaniel, B. T. and Cassell, B. G. (1981). Effects of embryo transfer on genetic change in dairy cattle. *J. Dairy Sci.* **64**, 2484–92.

Nicholas, F. W. (1979). The genetic implication of multiple ovulation and embryo transfer in small dairy herds. *30th Meeting Eur. Ass. Anim. Prod., Harrogate, England*, CG1.11.

—— and Smith, C. (1983). Increased rates of genetic change in dairy cattle by embryo transfer and splitting. *Anim. Prod.* **36**, 341–53.

Smith, C. (1981). Genetical and statistical aspects of physiological predictors of merit. *Scientific Workshop, Edinburgh*, Sept. 1981.

Trodahl, S. and Syrstad, O. (1977). *Ten years experience with cow indexing*. (28th Meeting of European Ass. Anim. Prod. Brussels, Belgium.) GC1.05. 4pp.

Van Vleck, L. D. (1981). Potential genetic impact of artificial insemination, sex selection, embryo transfer, cloning and selfing in dairy cattle. In *New Technologies in Animal Breeding* (eds B. G. Brackett, G. E. Seidel, Jr, and S. M. Seidel) pp. 221–42. Academic Press, London.

Walkley, J. R. W. and Smith, C. (1980). The use of physiological traits in genetic selection for litter size in sheep. *J. Reprod. Fert.* **59**, 83–8.

Discussion

It was enquired why the embryos transferred in the scheme would be frozen when they had 24-h survival *in vitro*. It was explained that the Danish experience was that there was a serious decrease in survival after 4 h and that the losses from freezing were less than those experienced by waiting for periods exceeding 4 h. It was pointed out that some calculations had appeared optimistic with eight ova on average obtained from each collection. It was explained that this was merely for calculation purposes and the scheme had been costed on the basis of 4.8 viable embryos per donor.

The physiological traits to be used were a battery of measurements and no selection would be practiced until the progeny test results became available and indicated the most useful. The most promising physiological predictors were thought to be insulin measurements in response to doses of propionate, blood urea, and measures of thyroid activity, but not growth hormone. Land pointed out some of the difficulties arising in choosing such predictors because in some instances the behaviour was like that of reproductive hormone. Addition of extra hormone could produce additional product but that observing the correlations within an untreated population, no correlations were observed.

Because the scheme was a joint Breed Association venture, high feeding standards would be adopted probably with *an lib* silage and concentrates. Individual feed intakes would be recorded every 14th day.

Measures to estimate genetic trends were discussed and it was explained that at present there would be no control in the nucleus herd but that performance tests and progeny tests of bulls from the scheme would be tested alongside other bulls. Smith pointed out the advantages of a deep frozen semen panel to establish a control, indicating that this could be used not only to measure trends in milk production (for which BLUP methods were was available) but also for other traits not routinely measured such as food efficiency.

In commercial progeny testing schemes for dairy cattle, there was a problem in the selection of bull dams, and the accuracy with which their breeding merit was estimated was not as high expected. The problems in small populations were equally difficult and the use of MOET was suggested. It was pointed out that the question of costs arose again particularly for small organizations. The cost element was also suggested as a reason why a MOET scheme for beef cattle which had been around for 10 years had not yet been used. It was pointed out that sexing of embryos would be useful in reducing the cost.

The question was raised as to how to deal with bulls with varying amounts of information some, for example, from a MOET test with feed efficiency values and others from progeny tests with estimates of breeding values for fertility. It was explained that in the Danish scheme there would be selection on total merit using a cow index with guestimates of relative values of different characteristics.

5

Somatic cell genetics and animal breeding

M. Förster

Abstract

Somatic cell genetics uses gene expression capacity under standardized in vitro *conditions. Somatic cell hybridization can be considered as an analogue to gamete fusion because it permits the incorporation of genetic material from diverse origins. Intraspecific cell hybridization allows the stabilization of highly specialized cell types and the conservation of their gene expression pattern by immortalization. Somatic cell hybrids are used for gene mapping and the production of monoclonal antibodies. Gene maps of domestic animals can be used for a better knowledge of linkage relationships and can help to isolate single genes. Monoclonal antibodies can be used for diagnosis and as protective antibodies against micro-organisms and parasites. Somatic cells can be used to create resistance genes and to study the molecular mechanisms of resistance genes in animals.*

Introduction

This paper focuses on the utilization of somatic cell genetics for the purpose of gene mapping, the production of monoclonal antibodies, and the creation of mammalian resistant cell mutants.

Cell hybridization

One of the most important single advances in somatic cell genetics has been the demonstration of somatic cell fusion. Often cell fusion is considered as an analogue to gamete fusion because it permits the incorporation of genetic material from diverse origins. Thus, it is possible to combine genes in various combinations, and investigate the interactions and the expression of such genes at their somatic level.

The fusion of the plasma membranes of adjacent cells results in the formation of hybrid cells. The fusion event can include two or more parental cells. The consequences are bi- or multinucleated cells. The proliferation capacity of such multinucleated cells is very low and most of them degenerate in only a few days. In most cases binucleated cells persist until the cells enter the next mitosis. Occasionally, such post-mitotic cells are mononucleated and possess the chromosomal material of both parental cell types. Such cells can proliferate often indefinitely and are termed somatic hybrid cells. Many cell lines adapted in *in vitro* conditions can be used for making viable hybrid cells. However, at least one parental cell type must be adapted to culture conditions for survival of the hybrids in culture, and for sustaining the proliferation and viability of the hybrid cells. Hybrids can be successfully formed either within or between species.

The spontaneous cell fusion frequency is very low, in the range of 1–5 × 10^{-6} for most cell types (Littlefield, 1966). Some viruses or inactivated viruses can facilitate the enhancement of cell fusion (Okada and Tadokoro 1963; Harris and Watkins 1965). In most current cell fusion experiments, polyethylene glycol (PEG) is used (Pontecorvo 1976). Now electrical field-mediated cell fusion opens new possibilities in cell fusion methodology (Zimmermann 1982).

A critical point in the use of somatic hybrid cells is the stability of the chromosomes. Intraspecific hybrids have the tendency to retain both parental chromosome sets (Yoshida and Ephrussi 1967). However, in interspecific cell hybrids, Weiss and Green (1967) found a strong tendency for elimination of one of the original parental chromosome set.

Gene mapping

Gene mapping by somatic cell hybrids is the classical use of interspecific hybrid cells. Gene assignment to specific chromosomes is based on random differential elimination of the chromosomes from the mapping species in combination with rodent cells. The absence of a marker is due to loss of the corresponding chromosome. This shall be discussed for the pig, because the pig has more assignments than other domestic mammals.

In specific fusion combinations of primary pig cell cultures (fibroblasts or leucocytes) and selectable permanent rodent cell lines, segregation of pig chromosomes occurs. In our case we fused pig fibroblasts and permanent mouse Rag cells. For the isolation of the hybrid cells HAT-selection was used as a selective system to eliminate the $HPRT^-$-Rag cells. In the remaining mixture of pig fibroblasts and pig-mouse cell hybrids, fibroblasts were grown from the hybrid cells in a few weeks. Some 27 different hybrid clones were thus isolated. In this procedure it is important that each clone arises from a single fusion event. Propagation of these clones allows the pig chromosomes to segregate randomly. This results in pig-mouse hybrid cell clones which differ randomly in their complement of pig chromosomes. For gene assignment to different pig chromosomes, phenotypic (in our case electrophoretic) and chromosomal data from a number of clones are analysed. If one pig enzyme always occurs with a certain pig chromosome and they are always lost together, it is likely that the structural gene for the enzyme is located on that specific pig chromosome.

We have now assigned genes to specific chromosomes in the pig: phospho-gluconate-mutase-1 (PGM-1) to chromosome 10, peptidase-B (PEP-B) to 11, lactatedehydrogenase-A (LDH-A) to 4 and -B (LDH-B) to 5, malic enzyme-1 (ME-1) to chromosome 1 or 17, malate dehydrogenase-1 (MDH-1) to 3, nucleoside phosphorylase (NP) and mannose phospho isomerase together to 7, and super oxide dismutase-1 (SOD-1) to chromosome 9 (Förster and Hecht 1984, 1985a, b). Gene assignment by somatic cell hybrid clones is also an efficient method for gene mapping by DNA recombination techniques.

Can gene mapping be useful in animal breeding? In theory a detailed knowledge of synteny and genetic linkage in domestic animals should be useful. An example in pig gene mapping done by family studies illustrates this. In the pig there exists a well known linkage group involving the genes for G blood group, phosphohexoseisomerase (PHI), halothane sensitivity, S inhibition factor, H blood group, serum post-albumin, 6-phospho gluconate dehydrogenase, SLA-complex, and J- and C blood groups. Fries *et al.* (1982) and Tikhonov *et al.* (1984) have assigned this linkage group to pig chromosome 15. In this case the close linkage of PHI and Hal could probably be used for the isolation of the gene for halothane sensitivity by recombinant DNA techniques. Here is an example of how a gene, studied by formal genetics, but unknown in its structure could be defined in its molecular basis.

Monoclonal antibodies

The most common example of the use of intraspecific somatic cell hybrids are monoclonal antibodies. One important differences between inter- and intra-specific cell hybrids is the chromosomal stability. In most cases intraspecific hybrid cells have a stable chromosome constitution. The ingenious idea in the production of monoclonal antibodies is the utilization of the capacity of cells to express genes in culture and to stabilize this gene activity by cell fusion for an unlimited period of cell culturing.

If single immune competent cells are allowed to proliferate after stimulation by a determinant, single definable antibody solutions can be obtained. By cell fusion of this antibody-producing cell with a B-lymphocyte cell having malignant growth characteristics and the capacity for unlimited division, immortalization of the antibody-producing cell takes place. For continuous antibody production, chromosomal stability of intraspecific hybrid cells is essential. Segregation of chromosomes from the genome of the determinant stimulated parental cell could disturb the antibody production capacity.

In *in vitro* populations of lymphoid cells, i.e. B lymphocytes, there will generally be more than one type stimulated by a given antigen. This is because antigens have more than one determinant and more than one B lymphocyte can recognize one determinant. Hence, each cell produces only one kind of specific antibody. Somatic cell genetics makes such antibody producing cells immortal by cell hybridization and available even on the single cell level.

The cloning of a single, immortal antibody producing hybrid cell leads to a cell clone producing a specific antibody. The resulting monoclonal antibody has highly specific binding characteristics to its antigenic determinant. The cell culture technique allows the continuous production of such monoclonal antisera.

Is there an application of hybridoma technology in animal breeding? In special cases monoclonal reagents can describe more precise family relationships in our animal populations than is possible by polyclonal antisera. We can also expect that the immunization procedure and the analytical quality can be more accurately standardized by the hybridoma technology. New areas of precise immuno-diagnosis become possible. Further applications are possible in animal parasitology and bacteriology. The detection of parasitic infections can be elucidated by monoclonal antibodies and hybridomas can produce specific protective antibodies directed against parasites (Yoshida *et al.* 1980). In the same manner Briles *et al.* (1981) have shown how mice can be protected against a certain type of *Pneumococcus* by monoclonal antibodies. In both cases, monoclonal antibodies are used as protective therapeutic substances. For animal breeding we can consider isolating antibody producing immunoglobulin genes from hybridomas. Furthermore, we can connect them with efficient gene regulation sequences and transfer these constructs into the genome of domestic animals by embryo micromanipulation. Of course, there are the expected problems of microbial mutation and the loss of resistance.

Mammalian cell mutants and resistance genes

In bacterial genetics cell mutants are very common. Somatic cell genetics allows us to produce mammalian cell mutants in a similar manner. The spontaneous mutation rate can be increased by exposure to mutagens. Early in somatic cell hybridization, cell mutants for the selection of the hybrid cells were needed. Littlefield (1964) presented TK- and HPRT-deficient mammalian cell mutants for HAT selection. Kao and Puck (1968) demonstrated the isolation of nutritional auxotrophic mammalian cell mutants.

Some years later Alt *et al.* (1978) reported on methotrexate-resistant mouse cells. In their paper a selective multiplication of the dihydrofolate reductase genes was demonstrated as the cause of the methotrexate resistance. A further report of methotrexate resistance was presented by Nunberg *et al.* (1978). In this case the amplification of the dihydrofolate reductase genes was confirmed and their localization on a single chromosome was shown. By means of *in situ* hybridization it could be demonstrated that the amplified DNA was localized to a homogeneously staining region (HSR) of a single chromosome. This HSR is an expanded region on a certain chromosome that is not present in sensitive cells of the same cell type. We can see that the resistance mechanism here is the amplification of a certain DNA segment and its integration in the chromosomal material. Such cells exhibit stable resistance.

In other cases, however, the loss of the same resistance has been due to loss of dihydrofolate reductase genes (Schimke *et al.* 1978). In such unstable cell lines so called 'double minutes' (DMs) can be observed in chromosome preparations. In these cell lines the number of dihydrofolate reductase gene copies was related to the number of DMs. In the absence of methotrexate, DMs and the supernumerary dihydrofolate reductase gene copies were lost (Brown *et al.* 1981). The excess of the enzyme diminished to the normal enzyme level.

These experiments demonstrate the ability of mammalian cells to amplify resistance genes, and give some answers to the phenomenon of stable and unstable genetic resistance. Two other cell mutants with interesting resistances might be mentioned here. These are cell mutants resistant to diptheria toxin (Gupta and Siminovitch 1978), and cadmium resistant mutants (Rugstad and Norseth 1978) as an example for heavy metal tolerance.

In somatic cell genetics we can use the capacity of somatic cells to produce genetic resistance by chance. Neither living domestic animals, because they are too few for a mutation experiment, nor the recombinant DNA techniques alone give us a comparable opportunity. For the creation of genetic resistance we need the inherent biological competence of the whole cell genome and a cell (individual) number greater than approximately 10^6. In the combination of somatic cell genetics and recombinant DNA techniques, the old idea of breeding for disease resistance in animals can be renewed. There can be no question but that we can learn a great deal about genetic resistance mechanisms by working on such research programmes.

References

Alt, F. W., Kellems, R. E., Bertino, J. R., and Schimke, R. T. (1978). Selective multiplication of dihydrofolate reductase genes in methotrexate-resistant variants of cultured murine cells. *J. Biol. Chem.* **253**, 1357.

Briles, D. E. Claflin, L. J., Schroer, K., Forman, C., Basta, P., Lehmeyer, J., and Benjamin, W. H., Jr (1981). The use of hybridoma antibodies to examine antibody mediated antimicrobial activities. In Hännerling et al. eds.: *Monoclonal antibodies and T-cell hybridomas*, Research Monographs in Immunology, (eds Hannerling *et al.*) Vol 3, 283.

Brown, P. C., Beverley, S. M., and Schimke, R. T. (1981). Relationship of amplified dihydrofolate reductase genes to double minute chromosomes in unstably resistant mouse fibroblast cell lines. *Mol. Cell. Biol.* **1**, 1077.

Förster, M., and Hecht, W. (1984). Some provisional gene assignment in pig. *Proc. 6th Eur. Colloq. Cytogenet. Domest. Anim.* **351**.

—— and —— (1985a). Genlokalisierungen für die Superoxid Dismutase (SOD-1), Nukleosid Phosphorylase (NP) und Mannose Phospho Isomerase (MPI) beim Schwein. *Zuchtungskunde* (in press).

—— and —— (1985b). Assignment of the genes for lactate dehydrogenase A and B in the pig to chromosome No 4 and 5 by somatic cell hybrids. *Ztschr. Tierz. Zuchtbiol.* (in press).

Fries, R., Dolf, G., and Stranzinger, G. (1982). Genkartierung bei landwirtschaftlichen Nutztieren: Gegenwärtiger Stand. *Schweiz. Landw. Monatshefte* **60**, 205.

Gupta, R. S. and Siminovitch, L. (1978). Isolation and characterization of mutants of human diploid fibroblasts resistant to diphtheria toxin. *Proc. Natl. Acad. Sci. U.S.A.* **75**, 3337.

Harris, H. and Watkins, J. F. (1965). Hybrid cells derived from mouse and man: Artificial heterokaryons of mammalian cells from different species. *Nature* **205**, 640.

Kao, F. T. and Puck, T. T. (1968). Genetics of somatic mammalian cells. VII Induction and isolation of nutritional mutants in Chinese hamster cells. *Proc. Natl. Acad. Sci. U.S.A.* **60**, 1275.·

Littlefield, J. W. (1964). Selection of hybrids from matings of fibroblasts *in vitro* and their presumed recombinants. *Science* **45**, 709.

—— (1966). The use of drug-resistant markers to study the hybridization of mouse fibroblasts. *Exp. Cell Res.* **41**, 190.

Nunberg, J. H. Kaufman, R. J. Schimke, R. T., Urlaub, G., and Chasin, L. A. (1978). Amplified dihydrofolate reductase genes are localized to a homogeneously staining region of a single chromosome in a methotrexate-resistant Chinese hamster ovary cell line. *Proc. Natl. Acad. Sci.* **75**, 5553.

Okada, Y. and Tadokoro. (1963). The distribution of cell fusion capacity among several cell strains or cells caused by HVJ. *Exp. Cell. Res.* **32**, 417.

Pontecorvo, G. (1976). Production of indefinitely multiplying mammalian somatic cell hybrids by polyethylene glycol (PEG) treatment. *Somatic Cell. Genet.* **1**, 397.

Rugstad, H. E. and Norseth, T. (1978). Cadmium resistance and content of cadmium-binding protein in two enzyme deficient mutants of mouse fibroblasts (L-cells). *Biochem. Pharmocol.* **27**, 647.

Schimke, R. T., Kaufmann, R. J., Alt, F., and Kellems, R. E. (1978). Gene amplification and drug resistance in cultured cells. *Science* **202**, 1051.

Tikhonov, V. N., Nikitin, S. V., Gorelov, I. G., Bobovich, V. E., and Astachova, N. M. (1984). Gene mapping of 10 loci in the pig chromosome No 15. *Proc. 6th Eur. Colloq. Cytogenet. Domest. Anim. Zurich*, p. 480.

Weiss, M. C. and Green, H. (1967). Human-mouse hybrid cell lines containing partial complements of human chromosomes and functioning human genes. *Proc. Natl. Acad. Sci. U.S.A.* **58**, 1104.

Yoshida, M. C. and Ephussi, B. (1967). Isolation and karyological characteristics of seven hybrids between somatic mouse cells *in vitro*. *J. Cell. Physiol.* **69**, 33.

——, Nassenzweig, R. S., Potocnjak, P., Nussenzweig, V., and Aikawa, M. (1980). Hybridoma produces protective antibodies against the sporozoic stage of malaria parasite. *Science* **207**, 71.

Zimmermann, U. (1982). Electric field-mediated fusion and related electrical phenomena. *Biochem. Biophys. Acta* **694**, 227.

Discussion

Sejrsen asked if there are any discernible patterns to the distribution of genes on chromosomes. Förster pointed out that there are too few markers in farm species, but that there is extensive conservation of linkage groups between species.

Bouquet queried the chromosomal assignment and linkage map presented for the pig using the data of Tikhonov *et al.* (1984) and Fries *et al.* (1982). Petersen also expressed doubt. Förster replied that these data were from family analysis and he could not say anything about the exactness of this result. Förster used the data only to illustrate the usefulness of a dense gene map in helping to isolate and characterize single genes by recombinant DNA techniques.

Archibald asked if somatic cell hybrids had been used to assign PHI and 6-PGD, but unfortunately these genes cannot be differentiated between mouse and pig by the electrophoresis used. Smith wondered how one can understand disease resistance in whole animals from studies in cell culture? Förster suggested that resistance genes can be found in cell cultures and the resistance mechanism can be studied in this system. Such genes found to be involved in resistance can be isolated and might be used in transgenic experiments. He pointed out that there are many examples of a genetic basis for disease or drug resistance in cell cultures.

6

Transgenics

Frank Gannon

Abstract

In transgenic animals novel molecularly cloned DNA is introduced into a freshly fertilized egg. In some cases, this DNA becomes an integral part of the animal's genome and is present in all cells of an animal that develops from that embryo. The major factors in this area of research which are discussed in this paper are the choice of animal, the choice of gene, and the choice of genetic environment which would optimize the possibilities of expressing the gene.

In agriculture, one very frequently knows the characteristics one would like to improve in an animal. Growth, milk yield, and efficiency of food conversion are obvious examples. Progress towards those goals has been impressive using classical selection methods, but it has been gradual. An alternative approach which may give an incremental improvement in predesignated characteristics now seems at hand. This is the method referred to as the generation of transgenic animals. Globally, this method involves the introduction of a molecularly cloned gene into the one-cell embryo such that, after reimplantation and development of the embryo, the novel information is present in all cells in the animal. A dramatic illustration which moved this idea from the realms of science fiction to publication in Nature came with the demonstration by Palmiter *et al.* (1982) that some transgenic mice which had acquired non-homologous growth hormone (GH) gene DNA did, in fact, grow to a size up to twice the size of non-transgenic litter mates. The main steps in the general method now used to obtain transgenic animals are summarized in Table 6.1.

Implicit in this method are four choices concerning (i) the animal, (ii) the gene which will be expressed, (iii) the DNA construction which will allow for a good and possibly induced level of gene expression, and (iv) the method of delivery of the DNA, i.e. micro-injection or viral vectors. For the purposes of this paper, the use of viral vectors will not be discussed.

Choice of animal

The choice of the animal is the result of different and sometimes conflicting factors: (a) the technical feasibility, (b) the availability of adequate numbers

Table 6.1. Transgenic animal method

1.	Select gene of interest.
2.	Prepare a DNA construction which should allow expression of that gene in animal cells.
3.	Collect freshly fertilized one-cell embryos.
4.	Micro-inject the DNA into the pronucleus.
5.	Reimplant the micro-injected eggs in pseudopregnant females.
6.	Analyse offspring.

and the physical characteristics of fertilized eggs, (c) the economic importance of the species, and (d) the compatibility of the species with the gene one wishes to express. At present, rumours abound about the existence of transgenic animals in species other than the mouse or frog. The species that have been mentioned include cows, pigs, rabbits, trout, chickens, and sheep. It seems clear that publications will soon follow showing that these species have, in fact, been successfully used to integrate novel DNA into their germ line. Further questions, such as whether the gene was expressed, the response of the animals and the heritability of the characteristics will be part of a major research effort in this area for some years to come. At this point, it seems inevitable that the technique will ultimately work in all species. The tactics which one employs to achieve that goal will be a matter for discussion. The extremes of the debate will be the advantages of laboratory animal systems due to the large number of relatively transparent eggs which mice and rabbits generate, the possible ease of using fish systems as large numbers of eggs that develop *ex utero* are available, and the difficulties of using large animals related to the paucity and opacity of their eggs, and the time it takes for them to breed which must be contrasted with their obvious economic importance.

The choice of gene

In the past few years, a very wide range of genes have been used in the generation of transgenic mice. These include growth hormone (GH) (Palmiter *et al.* 1982), thymidine kinase (Brinster *et al.* 1982), plasmid DNA (Gordon and Ruddle 1981), immunoglobulins (Brinster *et al.*, 1983), ovotransferrin (McKnight *et al.*, 1983), SV40 T antigen (Brinster *et al.* 1984), β-globin (Constantini and Lacy 1981; Stewart *et al.* 1982), insulin (Burki and Ullrich 1982) and growth hormone releasing factor (GRF) (Hammer *et al.* 1985).

From this, it is clear that the limitation in the future will be ingenuity and imagination of the researcher, and the physiological tolerance and limitations of the animal. A prime target of these experiments is increased growth and milk yield. Transgenic mice which over-expressed GH (Palmiter *et al.* 1982) or GRF (Hammer *et al.* 1985) responded with increased growth. This would indicate that both of these hormones might be considered to be present at subsaturation levels in mice. Exogenously administered GH also increased milk production in cattle (Peel *et al.* 1981) and GRF seems to have a similar effect (Brazeau *et al.* 1985). It would follow that domestic animals transgenic with respect to the GH gene should also have the potential to respond.

Beyond GH, the potential is vast. As an example, only one possibility will be presented. In collaboration with J.M. Sreenan and D. Headon, we propose to clone the Δ 12 desaturase gene and to micro-inject it with a view to making transgenic animals. If expressed, it should convert oleic acid to the polyunsaturated linoleic acid. The marketing and possible dietary value of this is obvious. The contributions to this meeting by R. Lathe and J. Mercier will probably give other specific examples of proposals to alter the heretofore natural products from domestic animals.

Beyond the realm of altering and improving the capabilities of domestic animals, transgenic animals also have a potential role directly in biotechnology. Proteins expressed in these animals could be very complex and would be correctly modified. Although the economics of this use of transgenic animals are not yet clear and although many proteins would have deleterious effects on the animal if over-expressed in the wrong tissues, this possibility is another driving force towards the development of this method.

Choice of expression system

The expression of any gene requires at least a promoter which is active in the organism under study. The actual amount of expression of the gene that will be obtained will depend on the strength of the promoter, the presence or otherwise of enhancers or blockers, and the chromosomal location of the gene, in addition to post-transcriptional factors. Because of the influence of the chromosomal locus, the simple reinsertion of an unaltered gene could remove the sequence from repressive and tissue specific influences, and allow a supraphysiological level of the protein to be present in an animal. Much research is currently ongoing to establish the best components and topology in the construction, but so far the approach can only be described as intelligent empiricism. In addition to the ability to express a protein at high levels, it will be important in some cases that the product is expressed in a tissue-specific manner. The effectiveness of tissue-specific expression in mouse experiments has been unpredictable to date although some specific examples, such as the expression of immunoglobulin gene in B lymphocytes (Storb *et al.* 1984; Rusconi and Kohler 1985), insulin gene-directed expression of SV40 large T antigen in B cells of the pancreas (Hanahan 1985), β-globin in erythroid cells (Chada *et al.* 1985) and myosin in skeletal muscle (Shani 1985) show that it is possible to obtain tissue specificity. In the context of domestic animals, an obvious choice for this purpose would be the casein genes which might limit expression to the mammary gland.

Conclusions

From this overview of transgenics, it is evident that much has already been achieved with this technique and the path for the full exploitation of the approach has been clearly indicated. The feasibility of the method, however, highlights our ignorance in many areas of molecular biology including the species specificity of proteins and the elements that control their expression, the limiting factors in many natural processes (for example, why did the transgenic mice not become 20 times bigger than normal), the heritability of genetic traits ('editing' has occurred in many transgenic mice), the factors which direct the chromosomal location of micro-injected DNA, and the influence this has on levels of expression. These questions together with environmental and economic ones will ensure that this area of research will remain of prime importance to Europe and the world for many years to come.

Acknowledgement

Support from the EEC Biomolecular Engineering Programme is gratefully acknowledged.

References

Brazeau, P., Pelletier, G., Petitclerc, D., Gaudreau, P., Lapierre, H., Couture, Y., and Morriset, J. (1985). Effects of human growth hormone releasing factor hGRF(1-44)NH$_2$ and analog hGRF(1-29)NH$_2$ on GH secretion in heifers and cows, and on milk production in lactating cows. In Quo vadis: *therapeutic agents produced by genetic engineering* (ed. A. Joyeaux *et al.*) pp. 64. Medsi/Gower Press.

Brinster, R. L., Chen, H. Y., Messing, A., van Dyke, T., Levine, A. J., and Palmiter, R. D. (1984). Transgenic mice harbouring SV40 T-antigen genes develop characteristic brain tumours. *Cell* **37**, 367-79.

——, Chen, H. Y., Warren, R., Sarthy, A., and Palmiter, R. D. (1982). Regulation of metallothionein-thymidine kinase fusion plasmids injected into mouse eggs. *Nature* **296**, 39-42.

——, Ritchie, K. A., Hammer, R. E., O'Brien, R. L., Arp, B., and Storb, U. (1983). Expression of a microinjected immunoglobulin gene in the spleen of transgenic mice. *Nature* **306**, 332-6.

Burki, K. and Ullrich, A., (1982). Transplantation of the human insulin gene into fertilized mouse eggs. *EMBO J.* **1**, 127-31.

Chada, K., Magram, J., Raphael, K., Radice, G., Lacy, E., and Constantini, F. (1985). Specific expression of a foreign β-globin gene in erythroid cells of transgenic mice. *Nature* **314**, 377-80.

Constantini, F. and Lacy, E. 1981. Introduction of a rabbit β-globin gene into the mouse germ line. *Nature* **294**, 92-4.

—— and —— (1982). Gene transfer into the mouse germ line *J. Cell Physiol.* Suppl. **1**, 219–26.

Gordon, J.W. and Ruddle, F.H. (1981). Integration and stable germ line transmission of genes injected into mouse pronuclei. Science **214**, 1244–6.

Hammer, R.E., Brinster, R.L., Rosenfeld, M.G., Evans, R.M., and Mayo, K.E. (1985). Expression of human growth hormone-releasing factor in transgenic mice results in increased somatic growth. *Nature* **315**, 413–6.

Hanahan, D. 1985. Heritable formation of pancreatic B-cell tumours in transgenic mice expression recombinant insulin/simian virus 40 oncogenes. *Nature* **315**, 115–22.

Lacy, E., Roberts, S., Evans, E.P., Burtenshaw, M.D., and Constantini, F.D. (1983). A foreign β-globin gene in transgenic mice: integration at abnormal chromosomal positions and expression in inappropriate tissues. *Cell* **34**, 343–58.

McKnight, G.S., Hammer, R.E., Kuanzel, E.A., and Brinster, R.L. (1983). Expression of the chicken transferrin gene in transgenic mice. *Cell* **34**, 335–41.

Palmiter, R.D., Brinster, R.L., Hammer, R.E., Trumbauer, M.E., Rosenfeld, M.G. Birnberg, N.C., and Evans, R.M. (1982). Dramatic growth of mice that develop from eggs microinjected with metallothionein-growth hormone fusion genes. *Nature* **300**, 611–5.

Peel, C.J., Bauman, D.E., Gorewit, R.C., and Sniffen, C.J. (1981). Effect of exogenous growth hormone on lactational performance in high yielding dairy cows. *J. Nutr.* **111**, 1662–71.

Rusconi, S. and Kohler, G. (1985). Transmission and expression of a specific pair of rearranged immunoglobulin μ and κ genes in a transgenic mouse line. *Nature* **314**, 330–4.

Shani, M. (1985). Tissue-specific expression of rat myosin light-chain 2 gene in transgenic mice. *Nature* **314**, 283–6.

Stewart, T.A., Wagner, E.F., and Mintz, B. (1982). Human β-globin gene sequences injected into mouse eggs, retained in adults and transmitted to progeny. *Science* **217**, 1046–8.

Storb, U., O'Brien, R.L., McMullen, M.D., Gollahon, K.A., and Brinster, R.L. (1984). High expression of cloned immunoglobulin κ gene in transgenic mice is restricted to B lymphocytes. *Nature* **310**, 234–41.

7

Reproductive efficiency in sheep

J.P. Hanrahan

Abstract

Variation in reproduction efficiency is largely a reflection of the number of lambs per parturition and the interval between parities. The former trait is a function of ovulation rate and embryo survival while the interval between parities is constrained by the existence of a period of seasonal anoestrus in most breeds. The literature on genetic variation in ovulation rate was summarized and shows that there is a considerable amount of variation both within and between breeds. Of special note is the evidence for a gene in Booroola Merino sheep with a large effect (~ 1.2 ova) on ovulation rate and the indication that genes with effects of similar magnitude are segregating in Javanese and Cambridge ewes. Recent evidence shows the existence of breed differences in embryo survival and there is also some evidence for breed differences in the distribution of ovulation rate. Calculations on various combinations of embryo survival and ovulation rate distributions suggest that a mean prolificacy of 1.9–2.0 with a high proportion of twin births (0.7) is feasible. Information on within breed variation in embryo survival and the duration of the breeding season is limited to repeatability estimates which suggest a low value (~ 0.1) for embryo survival and a medium value (0.3) for length of the breeding season.

Introduction

Female reproductive performance is one of the principal determinants of the efficiency of meat production especially for low fecundity species like cattle and sheep. A comprehensive index of reproductive efficiency would contain measures of output in terms of number, weight, and quality of offspring per unit time as well as the inputs utilized. A convenient first approximation for present purposes is the number of offspring produced per female per year. This index depends on the number born per parturition and on the interval between parities. The existence of a seasonal anoestrus period in most sheep breeds is a constraint on effective minimization of the interval between conceptions. Fertility at any particular exposure to males and the associated ovulation rate are influenced by the timing relative to the previous parturition and relative to the seasonal pattern of oestrus activity.

However, these complex interactions are secondary to the level of prolificacy per parturition which will be the main consideration of this paper since it is likely that a high prolificacy per lambing will be fundamental to any production system based on more than one lambing per year as well as being a major determinant of efficiency in flocks lambing once per year. Prolificacy (litter size) depends on ovulation rate and embryo survival and the main purpose of the present paper is to review information on the importance of genetic variation in these traits. Consideration will also be given to variability of litter size, since this can influence lamb survival and production costs, and seasonality of breeding.

Variation in ovulation rate

There is a large amount of variation among sheep breeds in ovulation rate with differences of almost threefold being recorded in some studies (Table 7.1). This variation has provided the means for genetic transformation of ewe prolificacy through cross-breeding with or without the development of composite breeds (Young *et al.* 1985; Boylan 1985; Razungles *et al.* 1985, Martin *et al.* 1981).

Table 7.1.   Some breed comparisons for ovulation rate

Source	Breed	Ovulation rate
Land *et al.* (1973)	Berrichon	1.13
	Romanov	2.57
Land *et al.* (1974)	Merino	1.06
	Finn	2.96
Hanrahan (1974)	Galway	1.57
	Finn	4.11
Bindon *et al.* (1985)	Merino	1.62
	Booroola	4.65
Lahlou-Kassi and	Timahdite	1.09
Marie (1985)	D'man	2.85

Information on the genetic contribution to within breed variation in ovulation rate is summarized in Table 7.2 in terms of the heritability of a single measurement. The relatively high heritability of ovulation rate in the Finn breed has been confirmed by divergent selection on ovulation rate at 18 months (Hanrahan 1982) which has generated a difference between the high and low lines equivalent to about 50 per cent of the control line mean at all ewe ages from 9–10 months to 4.5 years (Hanrahan 1984).

Piper and Bindon (1982) reported that the high prolificacy of the Booroola Merino was attributable to a gene with a large effect on litter size. Subsequent studies showed that the gene acted through ovulation rate.

Table 7.2. Estimates of the heritability of ovulation rate

Source	Breed	Heritability
Hanrahan and Quirke (1985)	Galway	0.32 ± 0.16
	Finn	0.50 ± 0.09
Ricordeau *et al.* (1982)	Romanov	0.24
Piper *et al.* (1980)	Merino	0.16 ± 0.07

The evidence for the effect of this gene (F) on ovulation rate (summarized by Piper *et al.* 1985) showed that the gene effect is about 1.2 ova, that the gene behaves additively, and that the size of the effect was independent of the background ovulation rate over the range 1.0–1.8. The gene effect is equivalent to approximately 2.5 phenotypic standard deviation units of a population with a mean ovulation rate of 1.6. Since the discovery of the F gene preliminary results on ovulation rate in a number of sheep populations suggest the segregation of a gene with an effect of similar magnitude to the Booroola gene. Thus, data on ovulation of the Cambridge breed presented a pattern which suggests the segregation of a gene with a large effect on ovulation rate. These data are summarized in Table 7.3 (Hanrahan and Owen 1985) and show a very high repeatability of ovulation rate coupled with a high co-efficient of variation — a pattern which strongly suggests that a gene with a large effect is segregating. Similar results have been observed for the variation of ovulation rate in Javanese ewes (Bradford *et al.* 1985).

Table 7.3. Ovulation rate in Cambridge ewes*

	Year	
	1983	1984
Mean	4.16	3.75
Range	1–13	2–12
Repeatability (53 ewes)	0.86	

*Hanrahan and Owen (1985).

Variation in embryo survival

The rate of embryo survival which is observed in ewes depends on the number of eggs entering the uterus. The probability of embryo survival declines in an essentially linear fashion as the number of embryos entering the uterus increases (Hanrahan 1980, and unpublished data). This pattern of decline is also evident when data from egg transfer experiments are examined and this suggests that the observed pattern is not genetically associated with ovulation rate *per se*. Because of the 'environmental' effect of ovulation rate on embryo survival, consideration of genetic differences in embryo survival must be at equivalent ovulation rates.

Hanrahan (1979, 1980) concluded that differences among breeds in embryo survival was not an important component of breed differences in litter size. However, information published recently suggests the existence of significant breed differences in embryo survival (Meyer *et al.* 1983; Ricordeau *et al.* 1982). In order to estimate the likely range among breeds with respect to embryo survival a number of estimates derived from various sources have been assembled in Table 7.4. This shows the probability of embryo loss estimated from data on multiple ovulating ewes which conceived using the expression given by Hanrahan (1980) which assumes that the number of surviving embryos follows a binomial distribution. The information on most breeds is limited to ewes shedding two ova. The data suggest a wide range of values for the probability of embryo loss when two ova are shed (0.06–0.26). The range of estimates for ewes shedding three ova is somewhat narrower (0.13–0.29). In all cases the probability of embryo mortality increases with the number of ova shed although the pattern is erratic probably due to sampling errors associated with small numbers of ewes in the most extreme corpora lutea category. If the estimates are averaged for the four groups with information on ewes shedding two to four ovulations the resulting probabilities are 0.14, 0.20, and 0.27. The data on the Romanov breed suggest a departure of 0.08 from these average values. Recent evidence on the Cambridge breed indicates that embryo survival is also significantly higher than the average values suggested by Hanrahan (1980) although further data are required.

Table 7.4.   Estimates of the probability of embryo loss

		No. corpora lutea			
Source	Breed	2	3	4	5
Hanrahan (1980)	Mixed	0.18	0.27	0.35	—
Kelly and Johnstone (1983)	Mixed	0.15	0.29	—	—
Meyer *et al.* (1983)	B. Leicester (BL)	0.24	—	—	—
	Romney	0.26	—	—	—
	BL × Romney	0.12	—	—	—
	Cheviot × Romney	0.17	—	—	—
	Booroola × Perendale	0.09	0.15	0.27	—
	Booroola × Romney	0.22	0.24	0.28	—
Ricordeau *et al.* (1982)	Romanov	0.06	0.13	0.18	0.20

The evidence on within breed variation in embryo survival is much more limited. The most direct evidence available comes from selection experiments for increased litter size. Results from four such experiments were summarized recently (Hanrahan and Quirke 1985) and these indicted that correlated changes in ovulation rate could account entirely for the observed responses in litter size. Available information on the repeatability of embryo survival within populations of ewes suggests a small permanent effect associated with the ewe (Hanrahan and Quirke 1985) with a repeatability estimate of 0.07. This indicates that genetic variation is likely to account for less than 0.05 of the variance in embryo survival and consequently an even smaller fraction of the variance in litter size.

The influence of three different embryo loss functions on mean litter size and on the proportion of twin births are given in Table 7.5 for flocks with different mean ovulation rates. The range in ovulation rate was 1–4 and the distributions were those observed in Galway, $\frac{1}{4}$ Finn $\times$ $\frac{3}{4}$ Galway, and Fingalway flocks belonging to the Agricultural Institute at Belclare. The probability of embryo loss for twin ovulating ewes (q_2) ranged from 0.06 to 0.18 with increases of 0.08 and 0.16 for ewes with three and four ovulations to give q_3 and q_4, respectively. The number of surviving embryos was assumed to follow a binomial distribution and the calculation of litter size excluded ewes with zero embryos. The results of these calculations show that the effect on litter size of varying embryo loss rate increases with mean ovulation rate. The proportion of twin births increased as the level of embryo loss declined for each mean ovulation rate. The maximum proportion of twins was 0.61 which is similar to the average value of 0.62 reported by Davis *et al.* (1983) for New Zealand flocks with litter size ranging from 1.7 to 2.3. The proportion of twin births depends on the distribution of ovulation rate as well as on the embryo loss function.

Table 7.5. Effect of varying embryo loss rate on average litter size ($\bar{x}$) and the proportion of twin births (P)

| Mean Ovulation rate | Embryo loss function* | | | | | |
| | 1 | | 2 | | 3 | |
	$\bar{X}$	P	$\bar{X}$	P	$\bar{X}$	P
1.69	1.46	0.40	1.53	0.45	1.59	0.50
2.23	1.82	0.53	1.93	0.57	2.04	0.61
2.41	1.90	0.53	2.03	0.56	2.15	0.59

*1: $q_1 = 0.18$, $q_2 = 0.26$, $q_3 = 0.34$;
2: $q_1 = 0.12$, $q_2 = 0.20$, $q_3 = 0.28$;
3: $q_1 = 0.06$, $q_2 = 0.14$, $q_3 = 0.22$;
where q_n = probability of embryo loss given that n eggs are shed.

Conception rate and pregnancy rate is usually lower in ewe lambs than in adult ewes and this is due in part to poor viability of fertilized eggs from ewe lambs as shown in egg-transfer studies (Quirke and Hanrahan 1977; McMillan and McDonald 1985). The reduced reproductive efficiency of ewe lambs shown in these experiments is not true for all breeds since Ricordeau *et al.* (1982) found no difference between embryo survival in ewe lambs and adult ewes of the Romanov breed following natural mating.

Distribution of ovulation rate

The distribution of ovulation rate will influence the distribution of litter size and hence the likely efficiency of reproduction since offspring survival declines substantially as birth rank increases (Bradford 1985; Hinch *et al.* 1983). Variation among breeds in the distribution of ovulation rate are difficult to quantify due to associated differences in mean ovulation rate. The Finn and Romanov breeds are both characterized by high mean ovulation rates and data on the distribution of the corpora lutea counts are summarized in Table 7.6. These data suggest that these breeds differ in the underlying distribution of ovulation rate. This difference combined with the low level of embryo loss in Romanov ewes (Table 7.4) would suggest that the Romanov would be a more effective source of genes for raising prolificacy in cross-bred populations than the Finn.

Table 7.6. Distribution of ovulation rate in young and adult Romanov and Finn ewes

No. of Corpora lutea	Romanov*		Finn†	
	$\bar{X} = 2.5$	$\bar{X} = 3.5$	$\bar{X} = 2.5$	$\bar{X} = 3.6$
1	0.02	0.00	0.03	0.02
2	0.49	0.05	0.51	0.14
3	0.44	0.55	0.36	0.38
4	0.04	0.31	0.09	0.26
5	0.01	0.08	0.01	0.13
6	0.00	0.01	0.00	0.07
No. observations	650	226	288	302
C.V.	0.21	0.20	0.29	0.30

*Ricordeau *et al.* (1982).
†J.P. Hanrahan (unpublished data).

Distribution of litter size

Bradford (1985) discussed the importance of the distribution of litter size and estimated the influence of greater uniformity of litter size on output per ewe.

The magnitude of the effect depended on the level of mortality assumed for triplet born lambs. The observed distribution of litter size is a reflection of the distribution of ovulation rate and the level of embryo survival. These two factors have been combined in calculations shown in Table 7.7. The derivation of the alternative ovulation rate distribution was based on information contained in Table 7.6 on this distribution in Finn and Romanov ewes. These distributions were used to derive threshold intervals (on an assumed underlying scale) between the different numbers of corpora lutea. These threshold intervals were then used to define two ovulation rate distributions with equal means. Application of two extreme embryo loss functions (Table 7.5) to the ovulation rates gave the litter size information. The calculations show that the distribution of ovulation rate has a minor effect on mean litter size with an expected favourable influence on the uniformity of litter size. The level of embryo loss had a large effect on mean litter size and also a substantial effect on the variability of litter size. These results can be used as a guide to interpretation of published data on litter size distribution. Thus, the data of Steine (1985) on litter size in four Norwegian breeds shows that for all breeds more than 66 per cent of ewes had twins and the mean litter size was at least 1.90. It is suggested that these breeds may have both a high embryo survival and a relatively uniform distribution of ovulation rate. Similar indications arise from the data on Texel sheep in the Netherlands published by Sharafeldin (1960). Information on ovulation rate in these breeds would allow some assessment of the role of the two components involved.

Breeding season

There is a considerable body of information on breed differences in the date of onset of breeding activity and on the duration of the breeding season. This information was reviewed recently (Quirke and Hanrahan 1984) and shows that breeds differ in both the date of onset and date of cessation of the breeding season. Evidence on the genetic basis for within breed variation is much less extensive. Purser (1972) and Land (1982) have shown that there is significant genetic variation within breeds for the date of onset of the breeding season, while Thrift *et al.* (1971) and Fahmy (1982) have reported heritability estimates of 0.25 ± 0.06 and 0.04 ± 0.07, respectively, for date of lambing. Dyrmundsson and Adalsteinsson (1980) have reported an association between coat colour genes and earliness of lambing in Icelandic sheep.

While genetic variation in the onset of seasonal breeding is evident there appears to be no information on the genetic relationship between date of onset and date of cessation of breeding activity. The basic objective for increased animal efficiency would be an extended breeding season. In this regard Quirke and Hanrahan (1984) reported a repeatability estimate of 0.30 ± 0.10 for the duration of anoestrus which suggests a modest heritability for this trait (or its complement the length of the breeding season).

Table 7.7. Influence of the distribution of ovulation rate and the level of embryo loss on litter size

	Distribution of the no. of corpora lutea			
	Model A		Model B	
1	0.21		0.11	
2	0.54		0.72	
3	0.23		0.17	
4	0.02		0.00	
Ovulation rate				
mean	2.06		2.06	
C.V.	0.35		0.25	
	Embryo loss function*			
	1	3	1	3
Litter size				
mean	1.69	1.89	1.71	1.91
C.V.	0.39	0.36	0.34	0.29
Proportion of				
singles	0.41	0.28	0.36	0.20
twins	0.48	0.56	0.57	0.69
triplets	0.11†	0.16†	0.07	0.11

*See Table 7.5 for definitions.
†Includes quadruplets.

Conclusion

In terms of the components of reproduction efficiency considered in this paper it is clear that there is an abundance of genetic variation for ovulation rate. Particularly relevant, perhaps, is the large effect on ovulation rate associated with a single locus in the Booroola Merino and the likelihood that single genes associated with effects of similar magnitude are present in Cambridge and Javanese breeds, and possibly also in Icelandic sheep (Jonmundsson and Adalsteinsson 1985). The mechanism whereby the Booroola gene effects an increase in ovulation rate is not completely understood, but feedback relationships between the ovarian follicle (inhibin) and the pituitary (FSH) are involved (Bindon *et al.* 1985). However, other ovarian factors may also be implicated (Cahill 1984) and the final resolution of the nature of the primary genetic lesion associated with the Booroola gene remains to be accomplished.

With regard to embryo survival there is now good evidence for breed differences, but the prospects for within breed selection on this trait appear limited since the available evidence on the repeatability of embryo survival suggests a value less than 0.1. However, existing breed differences could be exploited. In addition, there is evidence to show that in some breeds with a prolificacy of around 1.9 lambs per ewe the proportion of ewes producing twins is relatively high. This is most likely due to a combination of high embryo survival and a relatively uniform distribution of ovulation rate. Such breeds merit further investigation since more uniformity of litter size at high mean levels have an obvious advantage in terms of the efficiency of production.

Genetic variation among breeds in the components of the seasonality of breeding is evident, but prospects for genetic change within populations are less clear.

References

Bindon, B. M., Piper, L. R., Cummins, L. J., O'Shea, T., and Hilliard, M. A. (1985). Reproductive endocrinology of prolific sheep: studies of the Booroola Merino. In *Genetics of Reproduction in Sheep* (eds R. B. Land and D. W. Robinson) pp. 217-35. Butterworths, London.

Boylan, W. J. (1985). Crossbreeding for fecundity. In *Genetics of Reproduction in Sheep* (eds R. B. Land and D. W. Robinson) pp. 19-24. Butterworths, London.

Bradford, G. E. (1985). Selection for litter size. In *Genetics of Reproduction in Sheep* (eds. R. B. Land and D. W. Robinson pp. 3-18. Butterworths, London.

——, Quirke, J. F., Sitorus, P., Inounu, Tiesnamurti, B., Bell, F. L., Fletcher, I. C., and Torell, D. T. (1985). Reproduction in Javanese Sheep: evidence for a gene with large effect on ovulation rate and litter size. *J. Amin. Sci.* (In press).

Cahill, L. P. (1984). Folliculogenesis and ovulation rate in sheep. In *Reproduction in Sheep* (eds D. R. Lindsay and D. T. Pearce) pp. 92-8. Cambridge University Press, London.

Davis, G. H., Kelly, R. W., Hanrahan, J. P., and Rohloff, R. M. (1983). Distribution of litter sizes within flocks at different levels of fecundity. *Proc. N. Z. Soc. Anim. Prod.* **43**, 25-8.

Dyrmundsson, O. R. and Adalsteinsson, S. (1980). Coat color gene suppresses sexual activity in Icelandic sheep. *J. Heredity* **71**, 363-4.

Fahmy, M. H. (1982). Genetic parameters of date of lambing in DLS sheep. *Proceedings of the World Congress on Sheep and Beef Cattle Breeding* (eds R. A. Barton and W. C. Smith) **1**, 401-4. Dunmore Press Ltd., New Zealand.

Hanrahan, J. P. (1974). Crossbreeding studies involving Finnish Landrace and Galway sheep. In *Proceedings of the Working Symposium on Breed Evaluation and Crossing Experiments*, pp. 431-44. Zeist, Netherlands.

—— (1979). Genetic and phenotypic aspects of ovulation rate and fecundity in sheep. *Proceedings of the 21st British Poultry Breeders Roundtable, Glasgow*, pp. 1-21.

—— (1980). Ovulation rate as the selection criterion for litter size in sheep. *Proc. Aust. Soc. Anim. Prod.* **13**, 405-8.

—— (1982). Selection for increased ovulation rate, litter size and embryo survival. *Proc. 2nd Wld Congr. Genet. Appl. Livestock Prod.* **5**, 294-309.

—— (1984). Selection on ovulation rate in Finn sheep. *An Foras Taluntais, Animal Production Research Report (Dublin)*, pp. 71–72.

—— and Owen, J.B. (1985). Variation and repeatability of ovulation rate in Cambridge ewes. *Proc. Br. Soc. Anim. Prod.* paper no. 37.

—— and Quirke, J.F. (1985). Contribution of variation in ovulation rate and embryo survival to within breed variation in litter size. In *Genetics of Reproduction in Sheep* (eds R.B. Land and D.W. Robinson) pp. 193–201. Butterworths, London.

Hinch, G.N., Kelly, R.W., Owens, J.L., and Crosbie, S.F. (1983). Patterns of lamb survival in high fecundity Booroola flocks. *Proc. N.Z. Soc. Anim. Prod.* **43**, 29–32.

Jonmundsson, J.B. and Adalsteinsson, S. (1985). Single genes for fecundity in Icelandic sheep. In *Genetics of Reproduction in Sheep* (eds. R.B. Land and D.W. Robinson) pp. 159–68. Butterworths, London.

Kelly, R.W. and Johnstone, P.D. (1983). Influence of site of ovulation on the reproductive performance of ewes with one or two ovulations. *N.Z. J. Agric. Res.* **26**, 433–5.

Lahlou-Kassi, A. and Marie, M. (1985). Sexual and ovarian function of the D'man ewe. In *Genetics of Reproduction in Sheep* (eds R.B. Land and D.W. Robinson) pp. 245–60. Butterworths, London.

Land, R.B. (1982). Indicators of reproductive potential. *Proceedings of the World Congress on Sheep and Beef Cattle Breeding* (eds R.A. Barton and W.C. Smith), **1**, 365–73. Dunmore Press Ltd., New Zealand.

——, Pelletier, J., Thimonier, J., and Mauleon, P. (1973). A quantitative study of genetic differences in the incidence of oestrus, ovulation and plasma luteinising hormone concentration in the sheep. *J. Endocr.* **58**, 305–17.

——, Russell, W.S., and Donald, H.P. (1974). The litter size and fertility of Finnish Landrace and Tasmanian Merino sheep and their reciprocal crosses. *Anim. Prod.* **18**, 265–71.

Martin, T.G., Nicholson, D., Smith, C., and Sales, D.I. (1981). Phenotypic and genetic parameters for reproductive performance in a synthetic line of sheep. *J. Agric. Sci. Camb.* **96**, 107–13.

McMillan, W.H. and Mcdonald, M.F. (1985). Survival of fertilised ova from ewe lambs and adult ewes in the uteri of ewe lambs. *Anim. Reprod. Sci.* **8**, 235–40.

Meyer, H.H., Clarke, J.N., Harvey, T.G., and Malthus, I.C. (1983). Genetic variation in uterine efficiency and differential responses to increased ovulation rate in sheep. *Proc. N.Z. Soc. Anim. Prod.* **43**, 201–4.

Piper, L.R. and Bindon, B.M. (1982). Genetic segregation for fecundity in Booroola Merino sheep. *Proceedings at the World Congress on Sheep and Beef Cattle Breeding* (eds R.A. Barton and W.C. Smith) **1**, 395–400. Dunmore Press Ltd., New Zealand.

——, ——, Atkins, K.D., and McGuirk, B.J. (1980). Genetic variation in ovulation rate in Merino ewes aged 18 months. *Proc. Aust. Soc. Anim. Prod.* **13**, 409–12.

——, ——, and Davis, G.H. (1985). The single gene inheritance of the Booroola Merino. In *Genetics of Reproduction in Sheep* (eds R.B. Land and D.W. Robinson) pp. 115–25. Butterworths, London.

Purser, A. F. (1972). Variation in date of first oestrus among Welsh Mountain ewes. *Proc. Br. Soc. Anim. Prod.* **1**, 133. (*New Ser.*)

Quirke, J. F. and Hanrahan, J. P. (1977). Comparison of the survival in the uteri of adult ewes of cleaved ova from adult ewes and ewe lambs. *J. Reprod. Fert.* **51**, 487–9.

—— and —— (1985). Breed differences in the breeding season in sheep. In *Endocrine causes of seasonal and lactational anestrus in farm animals* (ed P. Ellendorff and F. Elsaesser) pp. 29–43. Martinus Nijhoff.

Razungles, J., Tchamitchian, L., Bibe, B., Lefevre, C., Brunel, J. C., and Ricordeau, G. (1985). The performance of Romanov crosses and their merits as a basis for selection. In *Genetics of Reproduction in Sheep* (eds R. B. Land and D. W. Robinson) pp. 39–45. Butterworths, London.

Ricordeau, G., Razungles, J., and Lajous, D. (1982). Heritability of ovulation rate and level of embryonic losses in Romanov breed. *Proc. 2nd Wld Congr. Gen. Appl. Livestock Prod.* VII, 591–5.

Sharafeldin, M. A. (1960). Factors affecting litter size in Texel sheep. *Meded. Landbouwhogeschool, Wageningen* **60**, 1–61.

Steine, T. (1985). Genetic studies of reproduction in Norwegian sheep. In *Genetics of Reproduction in Sheep* (eds R. B. Land and D. W. Robinson) pp. 47–54. Butterworths, London.

Thrift, F. A., Dutt, R. H., and Woolfold, P. G. (1971). Phenotypic response and time trends to date of birth selection in Southdown sheep. *J. Anim. Sci.,* **33**, 1216–23.

Young, L. D., Dickerson, G. E., and Fogarty, N. M. 1985. Evaluation and utilization of Finn sheep. In *Genetics of Reproduction in Sheep* (eds R. B. Land and D. W. Robinson) pp. 25–33. Butterworths, London.

Discussion

In response to questions, Hanrahan said that empty females were not included in the various tables he had shown. He did not know if the proposed Cambridge gene and the Booroola gene were the same as they had not yet been bred together. Identified animals with the Cambridge gene were not yet available.

Hill enquired if differences in variability associated with the different genotypes agreed with the recent changes in performance. He would have expected a very rapid response from increasing gene frequency. Hanrahan replied that the response had in fact been very rapid in the 1970's, but now that more homozygous Cambridge gene animals were being produced, their litter size would, if anything, be less than that of heterozygotes, so that future changes in litter sizes would be small. The response curve of litter size had in fact now flattened.

There was discussion on whether embryo survival depended on ova being shed from one or both ovaries, and the evidence to date was interpreted as arguable. The question was raised of why selection for ovulation rate in pigs had not produced any resulting increase in litter size. Hanrahan replied that the effect was basically the same as in sheep, with a curvilinear response. Embryonic survival seemed to depend in part on the age of the female, with uterine space being more severely limited in young females. Gibson enquired if all four populations with prolificacy genes had been subject to intense selection. This appeared to be so for the Cambridge and the

Booroola, but the Icelandic gene did not fit the same pattern and for the Javanese, there was no evidence for selection (although they incidentally appeared to have high embryonic survival). It was still not clear how the Booroola F gene acted and there was no clear cut gene product established. It was known that follicles produced less inhibin and there were differences in the recruitment of maturing follicles but the exact relationship was still unclear. Smidt enquired about the correlation between ovulation rate as expressed naturally and that induced by gonadotrophins. These correlations were positive, but most of them had been obtained between breeds and there was therefore hesitation about using them as physiological tests.

8

Double muscling in cattle

R. Hanset

Abstract

The results of an investigation on the genetic determination of the double-muscled condition in the Belgian Blue cattle breed are presented. Experimental and population data were used. Both led to the conclusion that a major gene, partially recessive (symbol mh*) is involved. Moreover, the double-muscled condition has been characterized from the anatomical, biochemical and cellular points of view. The next goals are the mapping, the cloning and the sequencing of this gene. The advent of the recombinant DNA technology makes these objectives attainable, the ultimate goal being the manipulation of the process of myogenesis that would be of great importance to the beef industry.*

Introduction

Extreme genetic stocks are normal candidates for the possible disclosure of major genes. The extreme muscled phenotype, acquired by the Belgian Blue cattle breed during the last 25 years, belonging obviously to this category, it was logically submitted to the appropriate genetic analysis. To carry out this investigation, two kinds of data were used: experimental and population data. On the way, data on the biological features of this condition have been collected. These will be presented and put in the perspective of a molecular genetic approach.

Inheritance

Experimental data

An experiment was carried out where F_1 cows all of the normal phenotype (born from the cross between Belgian Blue sires × Friesian cows) were back-crossed to Belgian Blue sires. The total weight of the most important muscles in the half carcase of calves slaughtered at the constant weight of 84 kg was the criterion for muscle development. In the offspring of this back-cross, the distribution of this criterion exhibits a clear bimodality corresponding to the segregation of two alleles in equal proportions. The two component distributions are referred to as BC_1 and BC_2, and their respective means are given in Table 8.1 (observed means). The symbol *mh* for muscular hypertrophy (and + for the normal allele) is proposed. Consequently, the three genotypes at the *mh* locus are written: *mh/mh*; *mh/+*; */+/+*.

At the same time, pure bred calves were produced from Friesian parents, from 'Belgian Blue' parents as well as F_1 calves. All the data available could be used together to estimate the gene effect. An additive model was fitted; the expectations corresponding to the five genetic types are given in Table 8.1 (Hanset and Michaux 1985a).

Table 8.1.   Expectations of the five genetic types (observed means, weighted least squares solutions, expected means)

Expectations	Observed means (kg)	Expected means (kg)
$D = \mu - m - a$	12.36 ($n = 5$)	12.36
$DM = \mu + m + a$	18.34 ($n = 30$)	18.38
$F_1 = \mu + d$	14.31 ($n = 7$)	14.50
$BC_1 = \mu + m/2 + d$	14.69 ($n = 28$)	14.64
$BC_2 = \mu + m/2 + a$	18.29 ($n = 32$)	18.25

Solution: $\mu = 15.369$; $m = 0.266$; $a = 2.746$; $d = -0.865$; residual standard deviation $= 0.791$. D = dairy; DM = double-muscled; F_1 = first cross DM × D; BC = back-cross.

In the expectations, *a* stands for half the difference between the genotypes *mh/mh* and *+/+*; *m* for half the difference due to other genes between the two breeds involved and *d* for the dominance deviation. These elements were estimated by solving the corresponding weighted least squares equations. The solutions are found in Table 8.1 and the relative positions of the three genotypes at the *mh* locus are illustrated by Figure 8.1 (Hanset and Michaux 1985a).

For the criterion used, the major gene for muscle hypertrophy behaves as a partially recessive, the heterozygous being closer to the homozygous normal; both exhibit the normal conformation. In practice, this means that the first crosses from dairy cows will have a better muscling and will be born easily since they don't exhibit the double-muscled conformation. Furthermore, any selection for better muscle development, in a population where the gene is present, would favour the heterozygotes.

Population data

In the commercial herds, bred by A.I., most of the cows are of the conventional type, but most of the A.I. bulls are of the double-muscled type, very few of the conventional type. Within these herds, segregation studies were therefore possible. The distribution of the phenotypes within the population could be analysed on two kinds of data: station and farm data (Hanset and Michaux 1985b).

The station data consist of 145 male calves sired by 11 fathers reared in

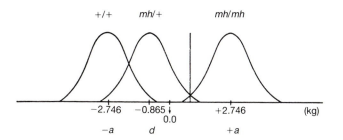

Fig. 8.1. The relative positions of the three genotypes (+/+, *mh*/+, *mh*/*mh*) at the *mh* locus according to the estimates (*a* and *d*) of the gene effects. The vertical line separates the normal conformation (at the left) from the double-muscled conformation (at the right).

station and slaughtered at 12 months of age. The following muscling criteria were recorded: dressing-out percentage (cold carcase 24 h after slaughter, liveweight recorded at the station 24 h before slaughter without any previous fasting), the lean and fat percentages in the 7th ribcut, the price per kg liveweight, the plasma creatinine level, and the content in creatine of the red blood cells (γ/100 ml). Their distributions were submitted to a normality test (test of Kolmogoroff–Smirnov) and as bimodality was revealed in each case, they were resolved into two component distributions, in the proportions p and q with means μ_1 and μ_2, and common variance, by a maximum likelihood procedure (Kaplan and Elston 1978). The results of this analysis are given in Table 8.2 (Hanset and Michaux 1985b).

Table 8.2. Test of normality (*D*) and the parameters of the component distributions (means μ_1 and μ_2, common variance σ^2, proportion q of double-muscled bulls, the difference $\mu_1 - \mu_2$ expressed in σ) of the distribution of six muscling criteria measured in Station on 1-year-old bulls ($n = 145$)

	D	μ_1	μ_2	σ^2	q	$(\mu_1 - \mu_2)/\sigma$
Price (B. F.)	<0.01	65.48	107.1	26.97	0.43	8.01
Dressing (%)	<0.01	59.55	64.18	1.20	0.45	4.23
Lean (%)	<0.01	57.48	69.23	10.50	0.41	3.63
Fat (%)	<0.025	12.99	22.89	10.55	0.52	3.05
Creatinine (γ/100 ml)	<0.01	1521.8	2084.9	29998	0.47	3.25
Creatine (γ/100 ml)	<0.01	1075.15	1934.7	55746	0.52	3.64

The magnitude of the differences between the means (several standard deviations) explains the bimodality of the distributions. This corresponds to the segregation, within the commercial population, of two distinct biological types: the conventional type and the double-muscled type.

The field data concerned 64 A. I. sires (59 double-muscled and 5 conventionals) and their 14 991 offspring of both sexes. The recorded characteristics were: the conformation of the parents and of the calf (two classes — double-muscled and conventional), a score of fleshiness on a scale running from 45 to 125 with steps of 5 units ascribed to each offspring when 1-year-old. The distributions of this score of fleshiness in the offspring of the four mating types (dm × dm; dm × conv; conv × dm; conv × conv) were analyzed. The tests of normality and the parameters of the component distributions are given in Table 8.3 (Hanset and Michaux 1985b).

Table 8.3. Test of normality (D) and the parameters of the component distributions [means μ_1 and μ_2, common variance σ^2, proportion q of double-muscled animals, the difference ($\mu_1 - \mu_2$) expressed in σ] of the distribution of the score of fleshiness at 12 months of age, for each mating type

Sire × dam	D	μ_1	μ_2	σ^2	q	$(\mu_1 - \mu_2)/\sigma$
dm × dm $n = 3326$	<0.01	74.76	102.83	73.55	86.11	3.27
dm × conv $n = 11\ 094$	<0.01	69.51	99.83	74.56	42.37	3.51
conv × dm $n = 62$	<0.01	71.92	97.86	60.66	49.16	3.33
conv × conv $n = 509$	<0.01	67.47	95.78	65.57	24.12	3.50

The segregation of two entities within a population, the shape of the distribution of the score of fleshiness for the different mating types, the similarity of the means of the component distributions across the matings, the proportions of double-muscled offspring, all these features support the hypothesis of a single major gene in agreement with what was shown in the first part.

The four mating types can then be written as follows (the conventional parents being considered as $mh/+$):

1. $mh/mh \times mh/mh$ exp. 100%; obs. 86.1%
2. $mh/mh \times mh/+$ exp. 50%; obs. 42.4%
3. $mh/+ \times mh/mh$ exp. 50%; obs. 49.2%
4. $mh/+ \times mh/+$ exp. 25%; obs. 24.1%

When comparing observations and expectations, recording errors and environmentally induced variation in the expression of the condition are to be taken into account. In mating dm × dm, for instance, cows could be recorded as double-muscled while they are really conventional and calves be considered as conventional while they are really double-muscled. Such errors of recording will result in a decrease of the expected percentage.

On the other hand, the possibility of a polygenic determination of the double-muscled character, considered as an all-or-nothing trait could be excluded since the finding of a break in the regression line of the proportions of double-muscled calves on the score of fleshiness of their dams, on a level with the antimode of the distribution of the score, is incompatible with this genetic model.

Therefore, as in the first part, it is concluded that a major gene is involved. Moreover, within the double-muscled class, a significant genetic variation in the degree of fleshiness could be demonstrated (Hanset and Michaux 1985b). This genetic variation, probably polygenic, may be exploited by selection.

Quantitative anatomy

The data collected in our back-cross experiment were used to compare the two segregating entities. The results of this comparison are given in Tables 8.4 and 8.5 separately by sex (R. Hanset and C. Michaux, unpublished data). In these tables, R stands for the ratio $(\text{dm/conv} - 1) \times 100$ and P for the level of significance of the difference between the two means.

Table 8.4. The main characteristics of the two biological types: double-muscled (dm) and conventional (conv) at the adjusted weight of 84 kg

	♂dm	♂conv	$R(\%)$	$P(\%)$	♀dm	♀conv	$R(\%)$	$P(\%)$
Gestation length (days)	285.9	279.2	+ 2.4	5.0	283.4	274.2	+ 3.3	1.0
Birth weight (kg)	49.4	42.4	+ 16.5	1.0	45.4	36.2	+ 25.4	0.1
Organs (kg)								
Skin	4.18	5.09	− 17.9	0.1	4.15	4.93	− 15.8	0.1
Liver	0.874	1.233	− 29.1	0.1	1.022	1.204	− 15.11	1.0
Heart	0.433	0.496	− 12.7	1.0	0.412	0.484	− 14.9	0.1
Lungs	0.693	0.884	− 21.6	0.1	0.706	0.862	− 18.1	0.1
Spleen	0.151	0.238	− 36.5	0.1	0.172	0.249	− 30.9	0.1
Thymus	0.123	0.269	− 54.3	0.1	0.155	0.260	− 40.4	0.1
Kidney (L.)	0.123	0.145	− 15.2	1.0	0.119	0.152	− 21.7	0.1
Kidney (R.)	0.115	0.141	− 18.4	1.0	0.117	0.154	− 24.0	0.1
Testes	0.013	0.018	− 28.3	1.0				
Mammary gland					0.125	0.175	− 28.8	10.0

Table 8.5.   The main characteristics of the two biological types: double-muscled (dm) and conventional (conv) at the adjusted weight of 84 kg

	♂dm	♂conv	R(%)	P(%)	♀dm	♀conv	R(%)	P(%)
Muscles (kg)								
(half-carcase)	18.51	14.91	+24.1	0.1	18.35	14.56	+26.0	0.1
Fat (kg)	0.44	0.74	−40.2	0.1	0.52	0.65	−20.1	10.0
Bones (kg)								
Scapulum	0.224	0.242	−7.4	1.0	0.208	0.225	−7.5	1.0
Humerus	0.450	0.489	−8.0	0.1	0.415	0.468	−11.3	0.1
Rad. Cub.	0.381	0.403	−5.5	5.0	0.342	0.369	−7.3	0.1
Femur	0.652	0.717	−9.1	0.1	0.609	0.683	−10.8	0.1
Tibia	0.455	0.478	−4.8	10.0	0.420	0.446	−5.8	5.0

The birth weight and the gestation length are significantly increased. Nevertheless, this larger size at birth is followed by a slightly lower size at subsequent ages. All the organs are hypotrophied, but to different extents, the spleen and the thymus being the most affected. Although the double-muscled male keeps its lower testis size through his life, at 1 year of age, the difference has dropped to 14 per cent (Michaux and Hanset 1981). The reduced size of the mammary gland was expected knowing the decreased milk production of double-muscled cows (− 15 to − 30 per cent) observed by Vissac *et al.* (1974).

The muscle hypertrophy is about 25 per cent, the weight of the adipose tissue is dramatically reduced in males, and the bones are the less affected. Muscle hypertrophy is not uniform; the superficial muscles are the most hypertrophied. These observations reveal a strong antagonism, a kind of competition during the development, between the muscular tissue and the adipose tissue and the different organs whose growth is delayed. The increase in birth weight due to the double-muscled condition is responsible for the calving problems. The general recourse to the caesarian operation has lifted this obstacle and opened the way to selection for more extreme animals.

Biochemical aspects

In double-muscled animals, the blood concentration in creatine is significantly lower (− 44 per cent) while the concentration in creatinine is significantly higher (+ 37 per cent) (see Table 8.2). Chemical analyses of muscle samples from double-muscled bulls, 1-year-old, have shown: (1) a similar DNA content; (2) a slightly higher nitrogen content; (3) a reduction by half of the intramuscular fat content; and (4) a reduction of 20–30 per cent of the hydroxyproline content (Hanset and Michaux 1982). The composition of growth, mainly its low fat content, explains the better efficiency of the double-muscled animals, at least on a high energy diet (Hanset *et al.* 1979). The content in hydroxyproline is an indirect measure of the importance of the connective tissue (collagen) which is, in part, responsible for the toughness of the meat. The 'double-muscled effect' in this respect seems to be more pronounced in muscles which normally have a higher hydroxyproline content. Differential hypertrophy of individual muscles, differential decrease of collagen content have led the Belgian butcher to develop a specific cutting system ('extensive cutting') which enhances the value of the carcase of double-muscled animals. Because of their size or because of their tenderness, some muscles, mainly in the forequarter, are commercialized as valuable cuts (Sonnet 1982).

The cellular level

The muscle hypertrophy syndrome is characterized by an increased number of fibres per muscle (hyperplasia instead of hypertrophy). The increase amounts to 18–24 per cent according to the muscle considered (Hanset *et al.* 1982). Furthermore, these animals have a higher percentage of white fibres (West 1974) and a higher proportion of branched terminal axons (Swatland 1973). This muscle fibre hyperplasia can be traced back to the early stages of primary fibre formation (Swatland and Kieffer 1974). The syndrome of muscle hypertrophy may be considered as originating from (1) accelerated mitosis at the multiplication stage (higher number of mitogenic receptors, higher concentration of growth factors in the surrounding medium at a very precise moment) or (2) delayed 'gene transition' from the nucleic acid metabolism (cell proliferation) to the synthesis of specific muscular proteins (cell differentiation) with, as a consequence, a longer persistency of mitogenic receptors on the cell surface. In this respect, it is worth mentioning that the degree of muscularity expressed as the ratio of the cubic root of the weight of the muscle semi-tendinosus on the length of the fetus reaches a maximum between the 6th and 7th month of gestation (Ansay 1976a) and that the small fetuses of the male sex exhibit a marked hypertrophy of the muscles of the neck as it will be again the case later in the adult bull (Ansay, 1976b).

The molecular genetic approach

It seems to us that the formal genetics of the double-muscled condition in cattle has reached an end, and that new approaches and new genetic tools are needed. Linkage and mapping are the next goals, but for this kind of investigation to be undertaken successfully requires a sufficient number of mapped polymorphic markers spread through the whole genome. If the distance between a gene of interest and a marker locus may not be greater than 10 cm, this defined 300 intervals of 10 cm assuming a total genome size of 3000 cm.

The 95 per cent probability that such an interval contains at least one marker gene implies an average of three genes per interval [by the Poisson distribution: exp $(-\mu)$ = 0.05] or a total of 900 (3 × 300) genes (Ruddle and Creagan 1975). Less stringent conditions, e.g. intervals of 20 cm and a 90 per cent probability would lead to a total of 345 genes (2.3 × 150).

However, the usefulness of marker loci in linkage studies depends on the number and frequency of their alleles. Botstein *et al.* (1980) have proposed the computation, for any marker locus, of a polymorphism information content (PIC) whose value tends to unity when the number of alleles with significant frequencies increases. The question then arises: is such a genetic knowledge available? Since the advent of the recombinant DNA technology, it has been shown that the genome is dense with sites of variability revealed by DNA fragment size polymorphism: the so-called RFLPs.

It is assumed that they are distributed widely and randomly throughout the genome. These RFLPs are detected using single copy sequences which have been cloned. For these probes there are five possible sources: (1) DNA segments corresponding to known genes; (2) synthesized oligodeoxyribonucleotides, but for that, the amino-acid sequence (or a part for it) of the coded protein must be known; (3) chromosome specific library; (4) cDNA library of a given organ or tissue; and (5) genomic library. These polymorphisms are mapped by *in situ* hybridization, by cell hybrid strategy or other cytogenetic methods. Coming back to the *mh* gene, 'gene-hunting' work can now be contemplated. Of course, the gene can be located anywhere on any of the 29 pairs and we can hope to get a hit like with Huntington's disease. On the other hand, if a few hundred markers were available, one or two of them would likely be close enough to the *mh* locus to show linkage. However, we could also try to determine the involvement of 'candidate' genes whose products play a role in growth and myogenesis, e.g. the genes for growth hormone, insulin, insulin-like growth factor I and II (somatomedins), oncogenes, genes for cell receptors, etc. In this respect, we can anticipate that the number of such cloned genes will continue to increase rapidly. Today, nobody knows when the *mh* gene will be mapped, cloned, sequenced, and compared to its normal counterpart. Very likely, this locus exerts its role in the regulation of myogenic cell proliferation and we know the relationship between the muscle fibre number and the muscle mass. Since myogenic cell proliferation is critical for the attainment of maximum muscle mass in livestock (Allen *et al.* 1979) it would be of paramount importance for the beef industry if we succeeded in manipulating the process of myogenesis at the right moment either during the prenatal period or during the post-natal period. If the basic function of the *mh* gene was unravelled, a great step towards this goal would have been made.

References

Allen, R.E., Merbel, R.A., and Young, R.B. (1979). Cellular aspects of muscle growth: myogenic cell proliferation. *J. Anim. Sci.* **49**, 115–27.

Ansay, M. (1976a). Existe-t-il, chez le foetus bovin, un âge de développement maximum de la musculature? *Ann. Méd. Vét.* **120**, 49–55.

—— (1976b). Développement anatomique du muscle de foetus bovin. Etude particulière de l'effet 'culard'. *Ann. Biol. Anim. Biochim. Biophys.* **16**, 655–73.

Botstein, D., White, R.L., Skolnik, M., and Davis, R.W. (1980). Construction of a genetic linkage map in man using restriction fragment length polymorphisms. *Am. J. Hum. Genet.* **32**, 314–31.

Hanset, R., Stasse, A., and Michaux, C. (1979). Feed intake and feed efficiency in double-muscled and conventional cattle. *Z. Tierzüchtg. Züchtgsbiol.* **96**, 260–9.

—— and Michaux, C. (1982). Creatine and creatinine levels in plasma, red cells and muscles as characteristics of double-muscled cattle. In *Muscle Hypertrophy of Genetic Origin* (eds. J.W.B. King and F. Menissier) pp. 237–56. Martinus Nijhof, The Hague.

——, Michaux, C., Dessy-Doize, C., and Burtonboy, G. (1982). Studies on the 7th rib cut in double-muscled and conventional cattle. Anatomical, histological and biochemical aspects. In *Muscle Hypertrophy of Genetic Origin* (eds J.W.B. King and F. Ménissier) pp. 341–9. Martinus Nijhoff, The Hague.

—— and Michaux, C., (1985a). On the genetic determinism of muscular hypertrophy in the Belgian White and Blue cattle breed. I. Experimental data. *Génét. Sél. Evol.* **17**, 359–68.

—— and —— (1985b). On the genetic determinism of muscular hypertrophy in the Belgian White and Blue cattle breed. II. Population data. *Génét. Sél. Evol.* **17**, 369–86.

Kaplan, E.B. and Elston, R.C. (1978). A subroutine package for Maximum Likelihood Estimation (*MaxLik*). Dept of Biostatistics. Univ. North Carolina, Chapel-Hill, Institute of Statistics. Mimeo Series no. 823.

Michaux, C. and Hanset, R. (1981). Sexual development of double-muscled and conventional bulls. I. Testicular growth. *Z. Tierzüchtg. Züchtgsbiol.* **98**, 29–37.

Ruddle, F.H. and Creagan, R.P. (1975). Parasexual approaches to the genetics of man. *Ann. Rev. Genet.* **9**, 407–86.

Sonnet, R. (1982). Analytical study on retail cuts from the double-muscled animal. In *Muscle Hypertrophy of Genetic Origin* (eds J.W.B. King and F. Ménissier) pp. 565–74. Martinus Nijhoff, The Hague.

Swatland, H.J. (1973). Innervation and genetically enlarged muscles from double-muscled cattle. *J. Anim. Sci.* **36**, 355.

—— and Kieffer, N.M. 1974. Fetal development of the double-muscled condition in cattle. *J. Anim. Sci.* **38**, 752–7.

Vissac, B., Perreau, B., Mauleon, P., and Ménissier, F. (1974). Etude du caractère culard. IX. Fertilité des femelles et aptitude maternelle. *Ann. Génét. Sél. Anim.* **6**, 35–48.

West, R.L. (1974). Red to white fibre ratio as an index of double-muscling in beef cattle. *J. Anim. Sci.* **38**, 1165–75.

Discussion

Hanset reassured Sellier that double-muscled cattle were not susceptible to halothane, so that the condition was different from that in pigs. Not all muscles were affected equally, but the superficial muscles showed more changes. A lively discussion ensued on the physical strength of double-muscled animals, but it emerged that muscular tissue was present in superabundance and was not required in full for muscular effort, but that there were heavy demands on the animal for good heart function. The muscle was innervated, but with a nerve supply which seemed to be more branched than in normal animals. In addition, there were more white muscle fibres than red ones although this did not provide an accurate enough test for the discrimination of heterozygotes from homozygotes. These differences in muscularity would probably lead to double-muscled individuals getting tired more quickly. A report that double-muscled animals showed more polyploidy in their cells had not been confirmed in any other subsequent report. Bouquet enquired about the role of minor genes in the expression of double-muscling, and it was thought that modifiers were present, but were dependent on the double-muscling gene for their expression. This led to a higher genetic variation in double-muscling animals although this might have been due to a scale effect.

Discussion of acceptability of double-muscled meat led to some disagreement about acceptability in different countries. In Belgium the double-muscled animals were favoured by butchers because they were able to make better use of some of the forequarter muscles that were less valuable in normal animals. Asked about the type of animals used by breeders, Hanset said that large scale breeders tended to use homozygous cows even though these gave more calving difficulties and less milk. Only small commercial farmers wanting milk used heterozygous cows (sometimes Friesian crosses), but wanted to get homozygous calves for sale. Discussion of the roles of sexing and embryo transfer led to Hanset's suggestion that the best way to use the technology would be to arrange for Holstein cows with high milk production to produce double-muscled calves.

9

Developments in estimating body composition and consequences for breeding programmes

Egbert Kanis

Abstract

Technical developments in estimating lean percentage in carcasses and in live animals are discussed. Classification of carcasses is at the moment mainly based on weight and subjective conformation scores. Recent development in evaluation of beef carcasses is directed at improvement of this scoring system. For assessment of pig carcasses automatic equipment with classification probes will be installed in slaughterhouses. This will lead to a decrease in RSD for prediction of lean percentage from about 3.3 per cent in the present EC classification scheme to less than 2.5 per cent. The corresponding changes in the payment system probably make it necessary to adjust the breeding objectives and their economic values.

New developments in assessment of live animals may be more drastic. With NMR and X-ray computerized tomography in principle it will be possible to predict the body composition of live animals of moderate size such as pigs and sheep to an accuracy of almost 100 per cent. For research work NMR and X-ray CT have great possibilities. Applications on a large scale in practical breeding programmes are not yet expected. Computer tomographs are too expensive, and their capacity and mobility are too low. More can be expected from ultrasound scanners. They can also be used in bigger animals such as cattle and are much cheaper than CT scanners. If they get smaller and portable the scanning can be done in the pen and the images can be analyzed later. The achieved prediction accuracy, however, is lower than with CT, but in pigs it is still high enough to exclude sib testing and to use all test places for boar performance testing.

Introduction

Together with milk and eggs, meat is one of the major animal products. Therefore, it is part of the breeding goal for almost all farm animals. Milk and eggs can be harvested more or less continuously during the life of the (female) animal. For these products it is relatively cheap and easy to obtain performance records from live animals. For 'harvesting' of meat the animals

81

have to be slaughtered and can not be used for breeding unless genetic material has been stored (frozen semen or embryos for instance). Estimation of breeding values for meat production is therefore to a great deal based on carcass observations or dissection results from sibs or progeny. Also indirect information from live animals such as conformation or ultrasonically measured backfat thickness is collected for breeding purposes. Both for the carcasses and for the live animals, new techniques are becoming available for estimating composition which can have great influences on breeding strategies.

This paper deals with the question why, what, and when we want to know (more) about body composition. A short review is given of current methods for estimating meat percentage *in vitro* and *in vivo*. Possible consequences of new technologies for breeding programmes are discussed.

Current methods

In many countries the evaluation of carcasses is mainly based on weight and visual assessment of conformation. Such assessments suffer from a number of disadvantages inherent in most systems of subjective judgement as reviewed by Kempster *et al.* (1984). In addition to visual judging in pig carcasses, fat depth measurements are often taken either with a ruler on the mid-line or with a probe. Sometimes also muscle depth is incorporated in the classification system as is the case in Denmark. Results from commercial carcass evaluation are rarely used for breeding purposes. For this aim often sibs from potential breeding animals are (partly) dissected.

In live animals, visual assessment and handling, together with body weight and breed are the most important sources of information for estimation of body composition for commercial purposes. For breeding work and research an increasing use of ultrasonic techniques can be observed. In pig breeding in particular ultrasonic equipment has been used for many years in many countries. The equipment available ranges from small portable A-mode instruments giving only fat depth to complex B-mode scanners capable of producing two-dimensional pictures of cross-sections through parts of the body. The principles of the use of ultrasound to predict percentage of lean are given by Simm (1983).

Why more?

Almost all farm animals are slaughtered at the end of their life, and the meat part of the carcass represents an economic value, even when the animals belong to breeds which are not used primarily for meat production. (In the Netherlands about 95 per cent of the total beef and veal production is from breeds primarily selected for milk production.) To estimate breeding values for meat production, accurate phenotypic information is needed about carcasses or rather live animals. However, most methods currently applied on a practical scale explain only a small proportion of the total variance in lean percentage. Kallweit and Averdunk (1984) concluded that the correlation between true lean content in pig carcasses and classification according to the widely used subjective EC classification scheme ranges from 0.57 to 0.67 with residual standard deviations ranging from 3.49 to 3.14 per cent lean. It is also known (Commission of the European Communities 1979) that great differences exist in application of the scheme between EC countries and that the classification is too coarse (steps of 5 per cent lean meat between two neighbouring classes). In live animals live weight and ultrasonically measured fat depths together explain about 50–55 per cent of the variance in carcass lean percentage for pigs (Kanis *et al.* 1986) and somewhat lower for cattle and sheep (Kempster *et al.* 1982).

What?

Heritable traits with a significant economic value now and in the future should be the objectives to select for. Since a payment system to the farmer will always be based on some type of meat content in the carcasses the estimation of meat percentage will be of major concern while the dimensions of muscles relative to the skeleton and meat quality are also important for the meat industry. From a consumers point of view it can be expected that meat tenderness, leanness, and taste will become more important in future (Steenkamp and Smidts 1985).

When?

Financial returns to the farmer are based on meat content in the carcasses at slaughter. Therefore, meat percentage at normal slaughter weight or age is most interesting. It can be expected that phenotypic information collected at that moment has the highest predictive value. However, if it were possible to predict later meat percentage in the young living animal, then selection could be done at an earlier stage, for instance before or during the first part of the fattening period, and selection intensity could be increased. Also for other, but breeding reasons it would be interesting to be able to follow the development of body composition in farm animals from birth to normal slaughterweight or to maturity.

New developments in carcass evaluation

A conclusion of Kempster *et al.* (1984) at the EAAP Satellite Symposium on 'Carcass Evaluation in Beef and Pork' was that visual assessments of beef carcasses are an essential element of 'the state of the art' and that they will be with us for some time to come. However, evidence is accumulating to indicate that some objective measurements may be complementary to the visual scores. Among these the use of video image analysis (Sørensen 1984) and automatic fat depth probes seem to be the most promising for practical use. It is doubtful however whether these methods become accurate enough for use in selection (for instance of progeny groups of A.I. bulls).

In pig carcass evaluation the use of objective measurements is much further developed than in cattle. The EC countries Denmark, Great Britain, and Ireland have used classification instruments in the slaughterhouses for several years while other countries have advanced plans to introduce classification devices in their abattoirs. This equipment is based on measurement of fat and muscle depths by insertion of probes at several predetermined points on the carcass. By means of prediction equations the percentage meat in the carcass is estimated directly. Research work in the Federal Republic of Germany (Averdunk *et al.* 1983) revealed correlations between instrument assessment (with FOM and HGP) and real meat content greater than 0.8 and RSD's less than 2.5 per cent meat. These figures are substantially better than obtained by the EC classification scheme. Furthermore, it will also be possible to evaluate some meat quality aspects, such as moisture content, pH, and colour with the same probes.

New developments in estimating body composition of live animals

Several developments in the field of *in vivo* estimation of body composition can be mentioned. A comprehensive survey is presented at the EC workshop in Langford. In this section the three currently most promising methods are briefly described. These are in ranking of increasing costs ultrasound scanning, X-ray computerized tomography, and Nuclear Magnetic Resonance computerized tomography.

Ultrasound

In Denmark the Danscanner equipment has been used for cattle, pigs, and sheep since 1976 (Busk 1984). This is a real-time ultrasound scanner with a multi-element transducer. In the U.S.A. the Scanogram has been developed which is based on a single ultrasound probe that is moved mechanically on a track across the back of the animal. Both machines provide two-dimensional scans of the back part of the animal which can be photographed and in which several measurements can be taken. The machines have been described and their results compared for cattle, pigs, and sheep by Kempster *et al.* (1982). They concluded that inclusion of areas of fat and muscle, over and above live weight and fat depth, in the prediction models, leads to a significant advantage in cattle. In pigs the improvements were smaller. Simm *et al.* (1983)

reported a comparison between Scanogram and Danscanner in cattle while an extensive review of the use of ultrasound to predict carcass composition in live cattle is given by Simm (1983). Recently, another ultrasonic machine has been tested by a pig breeding organization in the Netherlands (Molenaar 1984). It is based on the same principles as the Danscanner (real-time, B-mode, multi-element transducer) and equipped with a small unit for automatic image analysis. In addition to transversal scans longitudinal scans can also be taken. Some results from prediction of percent lean meat with models including backfat thickness, area of *M. longissimus dorsi* and feed intake per day for two lines of pigs are:

line	n	R^2	RSD
1	97	0.788	1.419
2	71	0.776	1.402
1 + 2	168	0.739	1.477

Molenaar (1984) concluded that use of ultrasound scanners in pig selection programmes can improve genetic progress significantly. Simm (1983), however, stated that marked improvements in the accuracy of these machines are unlikely since correlations between ultrasonic measurements and carcass composition are often as high as correlations between actual measurements on the carcass and total composition. Further improvements may come from techniques such as direct measurements of the velocity of ultrasound, which respond to inter and intramuscular fat as well as subcutaneous fat.

X-ray computerized tomography (CT)

X-ray CT has been developed for use in human medicine. Skjervold *et al.* (1981) reported the principles and the first application in animal production. The technique has been described by Vangen (1984) as the presentation of anatomical information from a cross-sectional plane of the body by computer synthesis of an image from X-ray transmission data obtained in many different directions through the plane under consideration. The image is constructed out of a matrix with so called CT values. A CT value is a function of the X-ray absorption in the corresponding spot of the body. Different tissues have different characteristic ranges of CT values. It is possible to include image areas (as in ultrasound techniques) and/or CT values in the prediction models. An important difference from ultrasound up to now is that X-ray CT is only usable for smaller farm animals as pigs, sheep, and goats. Recent results have been published in Langford at the CEC workshop and in The Hague at the EAAP conference (Allen and Vangen 1984; Vangen 1984; Sehested 1984[a,b]). Maximum R^2 value for predicting lean weight in the carcasses of 94 boars and 113 gilts was 0.621 (RSD = 0.95 kg) and for fat weight $R^2 = 0.813$ (RSD = 0.86 kg). These were considered as somewhat disappointing which was probably due to lack of precision in the dissection technique. Inclusion of CT means in the models was less accurate than inclusion of areas for predicting lean meat and only marginally better for prediction of fat weight. Higher accuracies were achieved in predicting chemical components of the bacon side (Allen and Vangen 1984). Vangen (1984) predicted protein (kg), fat (kg), and energy (MJ/kg) in the carcasses of 93 boars and 115 gilts. Inclusion of live weight, sex, and CT values from two scans in the model gave R^2 values for protein, fat, and energy of 0.89, 0.96, and 0.95, respectively, with corresponding RSD's of 0.503, 0.916, and 37.44. Sehested (1984[a,b]) scanned 293 ram lambs of the Dala sheep breed. He predicted the amount of protein (kg), fat (kg), and fat free lean (kg) in the carcasses. Models including live weight and CT values from four different cross-sections gave R^2 values of 0.93, 0.92, and 0.94, respectively, with corresponding RSD's of 0.122, 0.275, and 0.500.

It can be concluded from the Norwegian experiments that in particular the chemical composition in carcasses of pigs and sheep can be predicted to a very high accuracy. For prediction of percentage lean the results seem to be somewhat less satisfactory. There is need for a direct comparison between ultrasound scanning and X-ray CT.

NMR computerized tomography

The use of Nuclear Magnetic Resonance (NMR) to make images of cross-section 'slices' through the living body is a rather new technique, primarily developed for hospital use. An NMR imaging machine consists of a magnet with a central opening large enough to accommodate a patient (Wells 1984). This magnet applies a strong uniform magnetic field to the patient. This field tends to align the slightly magnetic hydrogen atoms (and some other atoms). The concentration, distribution, and properties of these protons in the body can be explored by measuring the electromagnetic signals which they radiate in response to slightly changing magnetic fields superimposed on the main field by electric currents passed through subsidiary coils enclosing the patient. Images can be generated corresponding to the intrinsic parameters, the so called relaxation times, of the protons in the tissues. As in X-ray CT these images or their underlying parameters can be analyzed. The signals emitted are, in fact, a product of the basic elements of the body, the atoms. Therefore, NMR CT is a more direct technique than X-ray CT which is based on X-ray absorption in body tissues. NMR CT is less sensitive to movements of the patient than X-ray CT and information can be collected from the whole of a volume of interest instead of only a single slice. It is also possible to measure flow or organ movements inside the body.

Until now not much work on the application of NMR CT in farm animals has been published. Groeneveld *et al.* (1984) reported results of four studies with X-ray CT and NMR CT in single animals or carcasses. They concluded that NMR differentiates soft tissue better than X-ray does. Deviating tissue such as tumours can be detected very well by NMR. Fuller *et al.* (1984) examined three pigs with a NMR imaging machine in Aberdeen and published several good quality images from cross-sections of live and dead animals.

Discussion and conclusions

Introduction of automatic classification equipment in abattoirs may have consequences for the traits in the breeding goal and their corresponding economic values so it seems likely that backfat thickness may not be a direct breeding objective any longer. Automatic classification of pig carcasses also gives new possibilities for using progeny information in selecting A. I. boars. An efficient identification system of pigs is then a basic requirement.

There is no doubt that with aid of modern scanning techniques in the near future it will be possible to predict the body composition of living animals with the same accuracy as present carcass dissection techniques. In addition, NMR CT and, to a lesser extent, X-ray CT can give information about the quality of the lean part. By repeated measurements during the life of animals it will also be possible to follow fat and protein deposition and to estimate maintenance requirements. For research purposes these techniques will certainly open new fields. For breeding purposes a costs–benefits analysis should be made. If the percentage lean meat can be determined *in vivo* with a sufficient accuracy, sib testing or progeny testing can be abolished and the test capacity can be used for performance testing, leading to higher selection intensities. Glodek (1984) suggests that a correlation of 0.7 between real body composition and live body measurement would allow more genetic gain from individual performance test than sib information or information from 25 progeny. If body composition can be estimated accurately in young animals this can lead to multi-stage selection strategies and so also to a higher expected genetic progress. Theoretically, this seems to be possible. In practice, however, many difficulties have to be overcome. The present CT machines are not suitable for scanning large animals such as cattle. Furthermore, exploitation of CT equipment is too expensive at present for most single breeding companies. Only on a national scale it might be economical to use CT in breeding programmes, provided that the flow of financial returns from the multiplier and fattening herds back to the nucleus is well organized. However, capacity and mobility of the present CT scanners are not high enough for efficient use on a national scale. Also the risk of disease will increase if many pigs are tested at the same place and spread over the country afterwards.

Ultrasonic scanners are probably somewhat less accurate in predicting lean meat percentage *in vivo* than CT, although a direct comparison is lacking.

Ultrasonic scanners, however, are far less expensive (about 65.000 Dutch guilders compared to 2-5 million for X-ray and NMR scanners). Much developmental work is still being done on ultrasonic scanning (especially for medical purposes) and it can be expeted that they will become smaller (even portable) and suitable for use in the environment of the animal. At the moment their prediction accuracy seems to be high enough to change breeding programmes completely to performance testing. In combination with computer software for image analysis the ultrasonic scanners probably have a better future in animal breeding programmes than computer tomographs.

References

Allen, P. and Vangen, O. (1984). X-ray tomography of pigs, some preliminary results. In *In vivo measurement of body composition in meat animals* (ed. D. Lister) pp. 52–66. Elsevier, London.

Averdunk, G., Reinhardt, F., Kallweit, E., Henning, M., Scheper, J., and Sack, E. (1983). Comparison of various grading devices for pig carcasses. *34th Ann. Me ting, Eur. Ass. Anim. Prod., Madrid.*

Busk, H. (1984). Improved Danscanner for cattle, pigs and sheep. In *In vivo measurement of body composition in meat animals* (ed. D. Lister) pp. 158–62. Elsevier, London.

Commission of the European Communities, (1979). *Development of uniform methods for pig carcass classification in the EC.* Information on Agriculture No. 70. Office for Official Publications of the EC, Luxembourg.

Fuller, M.F., Foster, M.A., and Hutchison, J.M.S. (1984). Nuclear Magnetic Resonance imaging of pigs. In *In vivo measurement of body composition in meat animals* (ed. D. Lister) pp. 123–33. Elsevier, London.

Glodek, P. (1984). The measurement of body composition. Opportunities and requirements in animal production. In *In vivo measurement of body composition in meat animals* (ed. D. Lister) pp. 8–20. Elsevier, London.

Groeneveld, E., Kallweit, E., Henning, M., and Pfau, A. (1984). Evaluation of body composition of live animals by X-ray and Nuclear Magnetic Resonance computed tomography. In *In vivo measurement of body composition in meat animals* (ed. D. Lister) pp. 84–8. Elsevier, London.

Kallweit, E. and Averdunk, G. (1984). Pig carcass classification and grading, perspectives for instrumental techniques. In *Carcass evaluation in beef and pork, opportunities and constraints* (ed. P. Walstra) pp. 45–54. IVO Schoonoord, Zeist, The Netherlands.

Kanis, E., Steen, H.A.M. van der, Roo K., de, and Groot, P.N. de (1986). Prediction of lean parts and carcass value from ultrasonic backfat measurements in live pigs. *Livest. Prod. Sci.* **14**, 55–64.

Kempster, A.J., Cuthbertson, A., and Harrington, G. (1982). *Carcass evaluation in livestock breeding, production and marketing.* Granada Publishing Ltd. London.

——, ——, and —— (1984). Beef carcass classification and grading methods, developments and perspectives. In *Carcass evaluation in beef and pork, opportunities and constraints* (ed. P. Walstra) pp. 21-9. IVO Schoonoord, Zeist, The Netherlands.

Molenaar, B. A. J. (1984). *De toepassing van een realtime echografie-methode bij het onderzoek van varkens.* Nieuw Dalland B. V., postbus 16, 5800 AA Venray, The Netherlands.

Sehested, ED. (1984a). Computerized tomography of sheep. In *In vivo measurement of body composition in meat animals* (ed D. Lister) pp. 67-74. Elsevier, London.

—— (1984b). Evaluation of carcass composition of live lambs based on computed tomography. *35th Ann. Meeting EAAP, The Hague.*

Simm, G. (1983). The use of ultrasound to predict the carcass composition of live cattle — a review. *Anim. Breeding Abs.* **51**, 853-75.

——, Alliston, J.C., and Sutherland, R.A. (1983). A comparison of live animal measurements for selecting lean beef sires. *Anim. Prod.* **37**, 211-9.

Skjervold, H., Grønseth, K., Vangen, O., and Evensen, A. (1981). *In vivo* estimation of body composition by computerized tomography. *Z. Tierzüchtg. Züchtgsbiol.* **98**, 77-9.

Sørensen, S.E. (1984). Possibilities for application of video image analysis in beef carcass classification. In *In vivo measurement of body composition in meat animals* (ed. D. Lister) pp. 113-21. Elsevier, London.

Steenkamp, J.E.B.M. and Smidts, A. (1985). Consumentengedrag vlees en vlees-waren. Vlees en Vleeswaren, mei 1985.

Vangen, O. (1984). Evaluation of carcass composition of live pigs based on computed tomography. *35th Ann. Meeting EAAP, The Hague.*

Wells, P.N.T. (1984). Introduction to imaging technology. In *In vivo measurement of body composition in meat animals* (ed. D. Lister) pp. 25-32. Elsevier, London.

Discussion

Some costs were questioned and Cameron suggested that cheaper ultrasonic instru-ments could be obtained, but it emerged that these would not include image analysis capabilities. The costs of NMR with helium cooling for metabolic measurements was agreed to be high, but thought by some that cheaper instruments for animals could be feasible, although it was pointed out that most of the push for new instruments came from medicine. For X-ray tomography, pigs have to be anaesthetized, but despite the risks, the group in Norway proved that it is possible to do this without loss of indi-vidual animals. With sheep it had not been found to be necessary to use anaesthetic. Smidt said that there should be a possibility of cheaper machines for use on carcasses in abattoirs although it emerged as uncertain as to whether individual scanning times would make this feasible. If images were not required then it was thought that fat to lean ratios might be obtained in less time.

It was pointed out that in addition to capital costs, the running costs of an operating team and statisticians needed to be considered. Also the low capacity in terms of number of animals scanned per unit of time and the manageability of the X-ray and NMR CT machines compared to ultrasonic equipment must be mentioned. It was questioned whether the extra costs by using CT above the newest ultrasound instru-ments are compensated by extra genetic improvements. Smith objected to this narrow consideration of costs arguing that real genetic improvements in a nucleus population would easily show the methods to be cost effective for the nation as a whole. Although companies might not be well organized to recoup their individual costs, it should be possible to convince national. fund-giving bodies of the utility of such instrumentation.

10
Novel products from livestock

R. Lathe, A. J. Clark, A. L. Archibald, J. O. Bishop,
P. Simons, and I. Wilmut

Abstract

Introduction of genes into the mouse germline is now a routine procedure.
Transferred genes are capable of developmental and tissue-specific expression
to give rise to functional products. Work with mouse is reviewed and extension
to farm animals is discussed. We propose that farm animals may be used as
vehicles for the production of human proteins of biomedical importance and
argue that, in many cases, the optimum system for expression will be the
mammary gland of the lactating ewe. Routes for achieving this objective are
discussed.

Introduction

Over the last few years techniques have been developed for introducing new
genetic material into the mammalian germline. Transfer of new genes into the
mouse is now a routine procedure and work is in progress to achieve this in
farm animals.

In 1980, Ruddle's group (Gordon *et al.* 1980) injected DNA into pronuclei
of fertilized mouse eggs and, following reimplantation into a suitable foster
mother, generated 'transgenic' mice in which the exogenous DNA could be
detected. Mintz (Wagner *et al.* 1981) and Brinster *et al.* (1981) further demonstrated that foreign DNA could be expressed in transgenic mice to give rise to
functional protein products. In parallel, Rubin and Spradling reported that
exogenous DNA could be introduced into the germline of the fruit fly,
Drosophila melanogaster (Spradling and Rubin 1982; Rubin and Spradling
1982).

Subsequent experiments reveal that foreign DNA can be transmitted from
transgenic mice to their progeny and be inherited in subsequent generations
(Wagner 1981; Costantini and Lacy 1981; Gordon and Ruddle, 1981;
Palmiter *et al.* 1982a; Stewart *et al.* 1982). Foreign DNA is capable of tissue
specific expression and foreign proteins produced from such exogenous
genes are functional (below). We extend here the view that novel gene combinations may similarly be introduced into livestock wherein they may determine the production of human proteins of biomedical importance. We

discuss below the techniques available for introducing DNA into animals, and also the choice of animal tissue for appropriate expression and harvesting of the target gene product.

Techniques for introducing DNA

Techniques for DNA transfer have been primarily developed in the mouse. Fertilized oocytes from newly mated female mice (1 day) are flushed from the oviduct and transferred to an appropriate incubation medium on a microscope slide. Each egg is held by gentle aspiration onto the tip of a micropipette, and the larger (male) pronucleus injected, using a microfine pipette and a micromanipulator, with about 2 picolitres of DNA solution containing some 200 copies of the appropriate construct. The eggs are then inserted into the oviducts of pseudo-pregnant mothers and allowed to develop. Survival and development is commonly 10–20 per cent, and the foreign gene sequences are found integrated into the DNA of a substantial proportion of the progeny.

Table 10.1 summarizes results obtained with mice to date. Some 15–20 per cent of mice born after the micro-injection of the eggs carry exogenous gene sequences integrated into their chromosomes. These mice, for the most part, transmit the integrated sequences to their progeny. Although in a small number of cases the incorporated DNA has been found to be rearranged, tandem integration of multiple intact copies of the injected DNA, at a small number of sites, seems to be the general rule (see Brinster *et al.* 1981).

The injection of mouse eggs is a straightforward procedure. In contrast, manipulation of the eggs of pig, sheep, and cattle has proved taxing. Early experiments with mouse eggs suggest that DNA injection into the pronucleus rather than cytoplasm increases the frequency of transgenic animals by some 20–30 fold. Mouse eggs are translucent and pronuclei are clearly visible. In the farm animals examined, without exception, ova are opaque due to the presence of numerous cytoplasmic vesicles, possibly fatty in nature. Physical treatments (such as centrifugation) have been used to visualize pronuclei, but such treatments reduce embryo viability (unpublished data).

An alternative approach may be to use viruses as vectors. Certain viruses, in particular retroviruses, have mechanisms which efficiently catalyse the integration of DNA copies into the host genome. However, the most evolved of the vectors available at present have important limitations. First, retroviral vectors have a limited capacity to transport foreign genetic material, and an upper limit of some 6 kb has been suggested (Cepko *et al.* 1984). This is possibly insufficient to permit the transduction of the large gene segments which may be required for tissue-specific expression. Secondly, the long terminal repeats of such vectors contain powerful transcription signals which

Table 10.1. Selected results obtained with transgenic mice

Authors (Construct)	Eggs reimplanted	DNA copies injected	Transgenic progeny	Germline Transmission	Expression	Site of Expression
Gordon et al. (1980) (PR-SV40 + hsvTK)	'sev 100s'	500–12000	2/173	nd	nd	
Wagner et al. (1981) (hBG, hsvTK)	ns	2500	5/33	nd	+	
Gordon and Ruddle (1981) [hIFN (alpha, cDNA)]	ns	10000	1/10	+	nd	
Costantini and Lacy (1981), Lacy et al. (1983) (raBG)	ns	1000	9/24	8/9	nd	muscle or testis
Palmiter et al. (1982b) (PR-mMT + rGH)	170	600	7/21	nd	+	primarily liver
Brinster et al. (1983) [mIgK (lit)]	192	440	6/11	4/4	+	spleen/lymphocytes
Swift et al. (1984) (rElastase type I)	320	200	7/37	7/7	5/5	pancreas
Grosschedl et al. (1984) [mIgM hev (r-r)]	284	50	5-13/23	3/4	4/5	lymphocytes
Shani (1985) (rMyosin lit-2)	ns	500	4/26	3/4	2/4	exclusively skel. muscle
Rusconi and Kohler (1985) [rIgM hev & lit (r-r)]	ns	20–100	5/13	4/5	4/4	B lymphocytes
Chada et al. (1985), Magram et al. (1985) (PR-mBG + hBG)	ns	10–30	10	+	4/10	erythroid cells
Hanahan (1985) (PR-rIns II + svT)	ns	20–150	5	5/5	+	exclusively pancreas (β-cell tumours)

h = human, hsv = herpes simplex virus, m = mouse, r = rat, ra = rabbit, sv = simian virus 40. BG = beta-globin, GH = growth hormone, IFN = interferon, Ig = immunoglobulin, Ins = insulin, PR = promoter, T = T antigen, TK = thymidine kinase. hev = heavy chain, lit = light chain. nd = not determined, ns = not specified, r-r = functionally rearranged.

may override signals determining the controlled expression of a transduced genetic segment.

A second possibility is to use embryonic cells capable of growing *in vitro* which can participate in the regeneration of a complete animal when injected into an early 'carrier' embryo and reimplanted (see e.g., Bradley *et al.* 1984). In principle, DNA can be introduced into these embryonic cells by standard calcium phosphate transfection techniques or micro-injection. However, to our knowledge, pluripotent embryocarcinoma cells have not yet been prepared from livestock.

At present the method of choice for generating transgenic livestock is injection into the pronuclei of fertilized eggs. The several-fold reduction in the viability of eggs resulting from centrifugation or other visualization procedures may be offset by the increased frequency of recovery of transgenic animals (> 10-fold) when pronuclear rather than cytoplasmic injection is employed.

Tissue specificity

In experiments with mice the expression of foreign DNA sequences is variable. Some animals express injected DNA with little or no tissue specificity or, indeed, in totally incorrect tissues (e.g., Costantini and Lacy 1981; Lacy *et al.* 1983). In other experiments, correct tissue-specific expression has been obtained (e.g. Storb *et al.* 1984; Swift *et al.* 1984; Shani 1985; Rusconi and Kohler 1985; Chada *et al.* 1985; Hanahan 1985; Magram *et al.* 1985). The general consensus is that correct tissue-specific expression of most if not all injected genes is now an achievable goal.

Transgenic animals and the production of pharmaceuticals

In recent years genes coding for a number of medically important human proteins have been isolated. These include the genes for insulin, growth hormone, plasminogen activator, and coagulation factors VIII and IX. At present these proteins are purified from blood and tissue, an expensive and laborious procedure which carries the risk of transmitting infectious agents such as that causing AIDS. The expression of these coding sequences in recombinant micro-organisms could in principle provide an alternative production method which avoids the risks associated with human products.

By and large, bacteria are unsatisfactory for the production of human proteins. Foreign proteins are often unstable in the bacterial host and the primary protein product is rarely, if ever, processed correctly. Expressing genes of interest in mammalian tissue culture cells has proved a viable strategy although batch fermentation of animal cells is expensive and technically demanding. The approach proposed here is to engineer domestic animals to act as vehicles for the production of human proteins of biomedical importance (Brinster *et al.* 1981; Lathe 1984).

Choice of tissue for expression

A number of factors must be taken into consideration with regard to the protein product and the tissue in which it is expressed. Sites wherein high levels of synthesis of the protein may be detrimental to the animal vector are obviously precluded, although this issue cannot be judged *a priori*.

Many proteins require extensive post-translational modifications in order to gain full activity. For example, coagulation factor IX (Christmas factor) requires gamma-carboxylation of a specific subset of glutamic acid residues (De Scipio and Davie 1979). Factor IX is synthesized in the liver and this tissue is, thus, proficient in performing this modification. It is, however, unclear to what extent such modifications are tissue specific or, alternatively, protein specific; few liver proteins are gamma-carboxylated. The possibility that factor IX possesses specific modification signals capable of dictating correct modifications in tissues other than the liver cannot be excluded. However, certain proteins may be correctly modified only if synthesized in certain tissues and, in this context, it may be necessary to tailor the site of expression to the requirements of the particular protein to be produced.

Other important considerations are the yield and ease of purification of the protein. Harvesting from a body fluid such as blood or milk (as opposed to a solid tissue) is desirable because (1) these routes of production are, by and large, renewable and (2), most proteins of biomedical importance are themselves secreted into body fluids.

A number of tissues secrete large amounts of protein into the bloodstream. The liver is an extremely active secretory organ and is the major source of many of the serum proteins. Some of these (e.g., albumin) are expresed at very high levels. The liver is capable of most forms of post-translational modification.

B-lymphocytes secrete large amounts of immunoglobulin into the blood. A number of workers (Table 10.1) have demonstrated that immunoglobulin genes, introduced into the germline of transgenic mice, are capable of relatively high levels of tissue specific expression. A possible advantage to using B-lymphocytes as the site of expression is that good cell culture systems (see e.g., Gillies *et al.* 1983) are available for testing DNA constructs prior to their introduction into the germline of the animal vector, to determine efficient tissue specific expression.

Although the large scale fractionation of blood proteins has been established for many years (Kohn *et al.* 1946) there are a number of disadvantages to the purification of proteins from this source. The coagulant properties of blood are a considerable hindrance to 'downstream processing' on a large scale. The presence of biologically active peptides (e.g., bradykinins) and antigenic molecules likely to co-purify with the product means that human proteins which are pharmaceutically acceptable will have to be extensively purified.

Isolation of human proteins from the milk of an animal vector may circumvent many of these problems. Milk is readily available in large quantities. It is well characterized biochemically and its protein composition is, relative to blood, quite simple. The milk protein genes are abundantly expressed and the mammary gland can carry out many forms of post-translational modification.

The mammary system

Mammary cells are adapted to high rates of secretion. They have specialized transport mechanisms for the efficient uptake of precursors from blood, and an extensive system of intracellular membranes (rough ER, Golgi, etc.) permitting high rates of protein synthesis, post-translational modification, and export from the cell.

The mammary gland is dependent on hormones for all aspects of its growth and function (for reviews see Topper and Freeman 1980; Forsyth, 1983). These mediate the striking changes that occur in the gland during pregnancy and lactation leading, for example in the ewe, to a near-doubling in total cell number as well as to changes in the proportions of the various cell types and to the terminal differentiation of secretory cells.

The mammary gland secretes a number of different proteins into the milk. There are qualitative and quantitative differences in the composition of milk from different species (Table 10.2), although a general distinction can be made between the caseins and the soluble (whey) proteins (see Jenness 1982). The major whey proteins are alpha-lactalbumin and beta-lactoglobulin.

The three major types of casein: alpha, beta, and kappa, are present in the milk of most mammalian species characterized. They serve to sequester calcium with which they are aggregated in the form of micelles. The function of beta-lactoglobulin (the major whey protein in ruminants) is unknown, although it appears to interact with kappa-casein (Brunner 1981). Alpha-lactalbumin, the second most abundant whey protein, is an essential co-factor in the conversion of glucose and galactose to lactose (Brew *et al.* 1968).

A major advantage of the mammary gland is that the major proteins are present in milk at high concentration (1–15 g/l). Milk protein genes are capable of correspondingly high rates of transcription. Given that DNA sequences associated with a milk protein gene can be artificially fused to and mediate the transcription of a chosen (exogenous) coding sequence, then high levels (several grams per litre) of the corresponding protein will be produced in the milk. When the normal levels in human blood of an appropriate target

Table 10.2. Protein composition of various milk.

g/litre	Bovine	Ovine	Murine	Human
Caseins			7	
alpha S1	10	12		
alpha S2	3.4	3.8		0.4
beta	10	16		3
kappa	3.9	4.6		1
Major whey proteins				
alpha-lactalbumin	1	0.8	trace	1.6
beta-lactoglobulin	3	2.8	no	no
whey acidic protein	no	no	2	no
Other whey proteins				
serum albumin	0.4	.	.	0.4
lysozyme	trace	.	.	0.4
lactoferrin	0.1	.	.	1.4
immunoglobulins	0.7	.	.	1.4

Data compiled from various sources.

protein are considered (for instance coagulation factor VIII is present at only some 0.1–0.5 mg/l) it becomes clear that a small number of livestock may be sufficient to satisfy our present needs for these proteins in the treatment of human disease.

Choice of experimental animal

The three common species of farm animal, pig, sheep, and cattle, each have advantages and disadvantages in relation to the objective of producing pharmaceuticals in milk (Table 10.3). In particular, the high milk yield obtained from cattle is offset both by our inability to manipulate the bovine embryo and the high cost of experiments involving cattle.

In a comparison between pig and sheep, transfer of embryos is nearly twice as expensive (per embryo) for the sheep as for the pig. The pig is not commonly regarded as a dairy animal and hence figures for milk yield are not readily available. Milk yield in sheep falls within the range of 1–3 l/day at peak season, depending on breed (but compare 10–30 l/day for cattle). It is of note that specialized equipment for harvesting milk from dairy sheep is available from commercial suppliers. We have therefore elected to pursue experiments with sheep rather than cattle or pigs.

Table 10.3. Comparison of different livestock for the purposes of generating of transgenic animals

	Pig	Sheep	Cattle
Availability of techniques for determining time of follicle maturation	Yes	(Yes)	(No)
No. of ovulations per animal without superovulation,	10	1–3	1
No. of ovulations per animal with superovulation	15–20	4–10	possibly 6
Visualization of pronuclei	(Yes)	(Yes)	(No)
In vitro culture of early embryo	(Yes)	(Yes)	No
Seasonal breeding	No	Yes	No
Relative cost of embryo transfer, including nominal cost of animals, per embryo	1.0	1.8	110

Gene combinations for germline transfer

As described in the previous sections, milk protein genes are abundantly expressed in the lactating mammary gland. In the ewe alpha S1 casein mRNA comprises about 30 per cent and beta-lactoglobulin about 8 per cent of (poly A +) messenger RNA (Mercier *et al.* 1985). Given that these transcripts originate from single copy genes, then these levels indicate high rates of transcription, although mRNA stabilization may also be involved. It is hoped to obtain similar levels of expression of the gene that is to be inserted into the ovine germline. This will be accomplished by linking it to those sequences associated with a milk protein gene which mediate the high levels of tissue specific expression.

In transgenic mice exhibiting high levels of tissue specific expression of a foreign gene (Table 10.1), the exogenous DNA comprises not only the structural gene, but also the 5′ and 3′ flanking sequences. There is considerable evidence that many important regulatory elements are located 5′ to the mRNA cap site (McKnight and Kingsbury 1982; Payvar *et al.* 1983; Renkawitz *et al.* 1984; Karin *et al.* 1984) although regulatory sequences, particularly those mediating tissue-specific expression, may often reside within the structural gene or even 3′ to it (Charnay *et al.* 1984; Gillies *et al.* 1983, Moore *et al.* 1985).

We are presently elaborating constructs for introduction into the ovine germline which may be expected to direct the efficient expression of particular proteins of biomedical importance in the mammary gland of the lactating ewe.

Discussion

In early experiments with transgenic mice bearing human growth hormone (GH) genes it was observed that the levels of GH recoverable from the sera of particular mice were considerably elevated. This led to the speculation that a production process might be developed in livestock (Brinster *et al.* 1981). Lathe (1985) extended this view and we now argue that the optimum biological production system is likely, for both technical and economic reasons, to be the mammary gland of the lactating ewe.

References

Bradley, A., Evans, M., Kaufman, M.H., and Robertson, E. (1984). Formation of germ-line chimaeras from embryo-derived teratocarcinoma cell lines. *Nature* **309**, 255-7.

Brew, K., Vanoman, T.C., and Hill, R.L. (1968). The role of α-lactalbumin and the A protein in lactose synthetase. A unique mechanism for the control of a biological reaction. *Proc. Natl. Acad. Sci. USA* **59**, 491-6.

Brinster, R.L., Chen, H.Y., Trumbauer, M., Senear, A.W., Warren, R., and Palmiter, R.D. (1981). Somatic expression of herpes thymidine kinase in mice following injection of a fusion gene in eggs. *Cell* **27**, 223-31.

——, Ritchie, K.A., Hammer, R.E., O'Brien, R.L., Arp., B., and Storb, U. (1983). Expression of a micro-injected immunoglobulin gene in the spleen of transgenic mice. *Nature* **306**, 332-6.

Brunner, J.R. (1981). Cow milk proteins: Twenty-five years of progress. *J. Dairy Sci.* **64**, 1038-54.

Cepko, C.L., Roberts, B.E., and Mulligan, R.C. (1984). Construction and application of a highly transmissible murine retrovirus shuttle vector. *Cell* **37**, 1053-62.

Chada, K., Magram, J., Raphael, K., Radice, G., Lacy, E., and Costantini, F. (1985). Specific expression of a foreign β-globin gene in erythroid cells of transgenic mice. *Nature* **314**, 377-80.

Charnay, P., Triesmann, R., Mellon, P., Chow, M., Axel, R., and Maniatis, T. (1984). Difference in human α and β-globin gene expression in mouse erythroleukemia cells. The role of intragenic sequences. *Cell* **38**, 251-63.

Costantini, F. and Lacy, E. (1981). Introduction of a rabbit β-globin gene into the mouse germ line. *Nature* **294**, 29-94.

De Scipio, R.G. and Davie, E.W. (1979). Characterization of protein 5, a. carboxyglutamic acid containing protein from bovine and human plasma. *Biochemistry* **18**, 899-904.

Forsyth, I.A. (1983). The endocrinology of lactation. In *The Biochemistry of Lactation* (ed. T.B. Mepham) pp. 309-49. Elsevier, Amsterdam.

Gillies, S.D., Morrison, S.L., Oli, V.T., and Tonegawa, S. (1983). Tissue-specific transcription enhancer element is located in the major intron of a re-arranged immunoglobulin heavy chain gene. *Cell* **33**, 717-28.

Gordon, J.W. and Ruddle, F.H. (1981). Integration and stable germ line transmission of genes injected into mouse pronuclei. *Science* **214**, 1244-6.

——, Scangos, G.A., Plotkin, D.J., Barbosa, J.A., and Ruddle, F.H. (1980). Genetic transformation of mouse embryos by microinjection of purified DNA. *Proc. Natl. Acad. Sci. USA* **77**, 7380–4.

Grosschedl, R., Weaver, D., Baltimore, D., and Costantini, F. (1984). Introduction of a μ immunoglobulin gene into the mouse germ line: Specific expression in lymphoid cells and synthesis of functional antibody. *Cell* **38**, 647–58.

Hanahan, D. (1985). Heritable formation of pancreatic β-cell tumours in transgenic mice expressing recombinant insulin/simian virus 40 oncogenes, *Nature* **315**, 115–22.

Jenness, R. (1982). Interspecies comparison of milk proteins. In: *Developments in Dairy Chemistry*, (ed. P.F. Fox) pp. 87–109. Elsevier, Amsterdam.

Karin, M., Haslinger, A., Holtgreve, H., Richards, R.I., Krauter, P., Westphal, H.M., and Beato, M. (1984). Characterization of DNA sequences through which cadmium and glucocorticoid hormones induce metallothionein-IIA Gene. *Nature* **308**, 513–8.

Kohn, E.J., Strong, L.E., Ewes, W.L., Mulford, D.J., Ashworth, J.N., Mellin, M., and Taylor, H.L. (1946). Preparation and properties of serum and plasma proteins. IV. A system for the separation into fractions of the protein and lipoprotein components of biological tissues and fluids. *J. Am. Chem. Soc.* **68**, 459–75.

Lacy, E., Roberts, S., Evans, E.P., Burtenshaw, M.D., and Costantini, F.D. (1983). A foreign β-globin gene in transgenic mice: integration at abnormal chromosomal positions and expression in inappropriate tissues. *Cell* **34**, 343–56.

Lathe, R. (1985). Molecular tailoring of the farm animal germline. Animal Breeding Research Organization 1985 Report, pp. 7–10. HMSO, Edinburgh.

McKnight, S.L. and Kingsbury, R. (1982). Transcriptional control signals of a eukaryotic protein-coding gene. *Science* **217**, 316–24.

Magram, J., Chada, K., and Costantini, F. (1985). Developmental regulation of a cloned adult β-globin gene in transgenic mice. *Nature* **315**, 338–40.

Mercier, J.C., Gaye, P., Soulier, S., Hue-Delahase, D., and Vilotte, J.L. (1985). Construction and Identification of recombinant Plasmids carrying DNAs coding for α-s_1, α-s_2, κ-casein and β-lactoglobulin. Nucleotide Sequence of αs_1 — Casein cDNA. *Biochimie* **67**, 959–71.

Moore, D.D., Marks, A.R., Buckley, D.I., Kapler, G., Payvar, F., and Goodman, H. (1985). The first intron of the human growth hormone gene contains a binding site for glucocorticoid receptor. *Proc. Natl. Acad. Sci. USA* **82**, 699–702.

Palmiter, R.D., Chen, R.Y. and Brinster, R.L. (1982a). Differential regulation of metallothionein-thymidine kinase fusion genes in transgenic mice and their offspring, *Cell* **29**, 701–10.

——, Brinster, R.L., Hammer, R.E., Trumbauer, M.E., Rosenfeld, M.G., Birnberg, N.C., and Evans, R.M. (1982b). Dramatic growth of mice that develop from eggs microinjected with metallothionein-growth hormone fusion genes. *Nature* **300**, 611–5.

Payvar, F., DeFranco, D., Firestone, G.C., Edgar, B., Wange, O., Okret, S., Gustafsson, J.A., and Yamamoto, K.R. (1983). Sequence specific binding of glucocorticoid receptor to MTV DNA and sites within and upstream of the transcription region. *Cell* **35**, 381–92.

Renkowitz, R., Schutze, G., Ahe, D., and Beato, M. 1984. Sequences in the promoter region of the chicken lysozyme gene required for steroid regulation and receptor binding. *Cell* **37**, 503–10.

Rubin, G. M. and Spradling, A. C. (1982). Genetic transformation of *Drosphila* with transposable element vectors. *Science* **218**, 348–53.

Rusconi, S. and Kohler, G. (1985). Transmission and expression of a specific pair of rearranged immunoglobulin μ and κ genes in a transgenic mouse line. *Nature* **314**, 330–4.

Shani, M. (1985). Tissue-specific expression of rat myosin light-chain 2 gene in transgenic mice. *Nature* **314**, 283–6.

Spradling, A. C. and Rubin, G. M. (1982). Transposition of cloned P elements into *Drosphilia* germ line chromosomes. *Science* **218**, 341–7.

Stewart, T. A., Wagner, E. F., and Mintz, B. (1982). Human beta-globin sequences injected into mouse eggs, retained in adults, and transmitted to progeny. *Science* **217**, 1046–8.

Storb, U., O'Brien, R. L., McMullen, M. D., Gollahon, K. A., and Brinster, R. L. (1984). High expression of cloned immunoglobulin κ gene in transgenic mice is restricted to B lymphocytes. *Nature* **310**, 238–41.

Swift, G. H., Hammer, R. E., MacDonald, R. J., and Brinster, R. L. 1984. Tissue-specific expression of the rat pancreatic elastase I gene in transgenic mice. *Cell* **38**, 639–46.

Topper, Y. J. and Freeman, C. S. (1980). Multiple hormone interactions in the development of the mammary gland. *Physiol. Rev.* **60**, 1049–106.

Wagner, E. F., Stewart, T. A., and Mintz, B. (1981). The human β-globin gene and a functional viral thymidine kinase gene in developing mice. *Proc. Natl. Acad. Sci. USA* **78**, 5016–20.

Wagner, T. (1981). Microinjection of a rabbit β-globin gene into zygotes and its subsequent expression in adult mice and their offspring. *Proc. Natl. Acad. Sci. USA* **78**, 6376–80.

Note added in press: since completion of this manuscript, Hammer *et al.* 1985, *Nature* **315**, 680–3) report the generation of transgenic livestock (sheep and pigs).

Discussion

Hill commented that the proposed use of farm species to produce novel protein products is not what people regard as genetic manipulation, i.e. the manipulation of genes affecting performance. Lathe suggested that the manipulation of commercial traits is poorly understood, e.g. the insertion of the human growth hormone gene into mice causes the females to be sterile. This side-effect would be very undesirable in farm species. There is a real dichotomy between the strategies of improving commercial traits or producing novel products. In the latter case we know the genes of interest and have the technology to manipulate them; we know little of the genes affecting production traits. Booman asked if we must know about all the genes affecting production traits? Lathe replied no, but pointed out that for a trait like twinning in cattle we know very little about any genes involved.

King asked if the results with mice transgenic for growth hormone were not dramatic enough to justify doing this in farm species. Lathe said there are a number of reasons for not doing this particular experiment in Edinburgh.

1. Humans with high growth hormone levels are medically compromised so animals might also be compromised.
2. Large cattle breeds already exist, but are not always used.
3. Fast growing, efficient animals do not show elevated growth hormone levels, e.g. broiler chickens.
4. Current experiments in other laboratories with pigs have failed to produce fast-growing animals.
5. Other laboratories are attempting the same experiments with growth hormone.

Hill asked who had done the experiments on pigs and whether growth hormone releasing factor (used in another experiment) overcame the problems mentioned with growth hormone. Lathe referred him to a paper in the current issue of *Nature* where Brinster *et al.* had just published the work on pigs. A lot of interest is now being shown in manipulating growth hormone releasing factor but its effects cannot be predicted. Gibson said it is not clear that we want to produce larger cows, and Lathe agreed that it would be difficult to assess the value of extra growth even if we succeeded in producing it.

Mercier pointed out that the rabbit can produce up to 1 litre of milk per day, and queried whether we should do experiments to produce high value proteins in livestock. Lathe agreed that the cost were much higher in livestock and that the use of rabbits should be explored as an alternative. Livestock have advantages in producing these novel products over yeast or bacteria. Most proteins have to be processed and modified after translation and yeasts and bacteria do not have the capacity to produce the final, active forms of most mammalian proteins. To produce these proteins from cultures of mammalian cells has a high recurrent cost whereas having some cows producing these proteins in milk would be cheap. Smith asked how many cows would be required to fulfil the national need for such high value proteins. It was thought that perhaps 10–20 would suffice.

King remarked that we usually think of genetic manipulation as making things work better but perhaps we wish to stop some processes, e.g. to stop the synthesis of cholesterol, or certain fats. Can anti-sense DNA switch genes off? Lathe agreed that the production of anti-sense RNA to block gene expression does seem to work in cell culture, and might be adopted to animals.

Hill asked if we can expect to have transgenic sheep of practical value within a few years or whether there are still problems to overcome. Lathe thought there are still many problems, but even if in 2 years there are animals producing new products, it will then take 10 years to develop them.

Sejrsen asked whether, in view of the possible disadvantages of continuous expression of introduced genes, is it not important to be able to switch genes on and off? Lathe agreed and pointed out that for example with a metallothionein promoter, dietary levels of zinc or cadmium might be used to switch gene expression on and off. Sejrsen also pointed out that the failure of added growth hormone in transgenic pigs to induce extra growth gives us valuable information about their physiology and Lathe agreed. Finally, Smith asked if it was envisaged doing the genetic manipulation serially, i.e. producing a number of products from the same animals? Lathe thought that the production of different products may be incompatible in one animal and the numbers required for each product are small.

11

Whole genome transfer in mammals

R. H. Lovell-Badge and J. R. Mann

Abstract

Recent advances in technology have allowed experiments involving the transfer of nuclei between cells of early mammalian embryos to be carried out efficiently and with little loss of viability. We review here results obtained with these techniques on some aspects of mouse embryogenesis, and discuss their significance to possible animal breeding strategies. Some maternal effect mutations producing hybrid incompatibility may be overcome by transferring pronuclei to normal egg cytoplasm. However, in at least one case, the defect rests with the pronuclei and may be attributed to the phenomenon of 'imprinting' when maternally and paternally inherited genomes behave differently. Imprinting also rules out the possibility of obtaining uniparental offspring as both maternal and paternal pronuclei are essential for normal development. Attempts to clone mice by transferring later stage nuclei into enucleated one-cell embryos have been unsuccessful. However, while imprinting may prevent the use of cells from adults as donors, the failure to date may simply be due to the wrong choice of recipient cytoplasm.

Introduction

The techniques of nuclear transfer have been very important for our understanding of the roles of nucleus and cytoplasm in the maintenance of the differentiated state. Experiments with amphibian embryos, conducted over a period of more than 30 years, have shown that nuclei from early embryos and even some nuclei from tadpole stages can support development to normal adults when transferred to enucleated eggs (Briggs and King 1952; Gurdon 1962a,b). Nuclei from adult cells seem to be more restricted as the latest stage obtained has been the feeding tadpole (Di Berardino *et al.* 1984). Nevertheless, it is clear from these experiments that the differentiated cell nucleus can be 'reprogrammed' after transfer into egg cytoplasm. It has, until recently, been very difficult to carry out similar studies on mammalian embryos, and it was not clear to what extent the mechanisms operating to control developmental potential or maintain differentiation are the same. New techniques of nuclear transfer have now made such studies possible, however, and we briefly review here some of the results from a number of laboratories including our own. In particular we shall consider the possibility of cloning animals as this is the aspect likely to have most impact on strategies for animal breeding.

Methodology

The techniques that have been successfully used with amphibian embryos and oocytes have generally involved puncture of the egg membrane to remove and introduce nuclei (although UV irradiation has been used to functionally 'enucleate'). Mammalian eggs are much smaller and have comparatively large nuclei. Puncture of the plasma membrane with pipettes large enough to transfer nuclei is often too disruptive and many embryos lyse. Results have been obtained using such techniques on mouse embryos (e.g. Illmensee and Hoppe 1981; Modlinski 1981), but their interpretation has sometimes been difficult due to low experimental numbers, the inability to carry out reciprocal transfers, and general problems in repeating the experiments. In 1983, McGrath and Solter described their technique for nuclear transfer which can have a success rate of up to essentially 100 per cent, when done between pronuclear stages, as it does not involve penetration of the egg membrane at any point. In the presence of cytochalasin and colcemid (or nocadazole) the egg cytoplasm is very fluid, and portions of cytoplasm, with or without nuclei, can be removed by pressing down on top of a zona free egg with a glass needle (Surani and Barton 1983). Using a bevelled, but fairly blunt pipette McGrath and Solter (1983) showed that it was possible to penetrate the zona of eggs cultured in these inhibitors and position the pipette over a pronucleus without breaking the membrane. One or both pronuclei, surrounded by a small amount of cytoplasm and plasma membrane are then sucked into the pipette. When the pipette is withdrawn the membrane will pinch off and seal leaving behind an intact cytoplast. The karyoplast within the pipette or a whole cell can then be fused to a previously enucleated cytoplast with inactivated Sendai virus. Fusion normally takes place rapidly and the embryos can be removed from the cytoskeletal inhibitors and either cultured *in vitro* or reimplanted into a foster mother with essentially no loss of viability. The only disadvantage of this technique is that a small amount of cytoplasm and membrane are transferred with the nucleus or pronuclei, but this can often be overcome by appropriate controls.

The method should be generally applicable to most cell types and to species other than mouse, although some modifications may be necessary. For example, levels of cytoskeletal inhibitors may be critical and alternative methods of fusion may be required. Both problems were encountered with hamster eggs (R. J. Mann and M. Orsini, unpublished observations) and some mouse embryonic cell types fuse poorly (R. J. Mann and R. H. Lovell-Badge, unpublished observations). A suitable alternative to Sendai virus may be provided by the technique of electrofusion (Zimmermann and Vienken

1982). Another problem that may be encountered is difficulty in visualizing the nuclei or pronuclei. The eggs of many domestic species, such as cattle, sheep, and pigs, contain numerous granules and lipid droplets, and are opaque. However, it has recently been shown that the nuclei in these eggs may be revealed by centrifugation and/or interference contrast microscopy with no effect on viability (Wall *et al.* 1985; Hammer *et al.* 1985). It may also be possible to use DNA specific fluorochromes with low levels of UV light and image intensifiers to visualize nuclei or higher levels of UV to functionally enucleate.

Observations

These nuclear transfer techniques have been used in at least three areas of research: to investigate (i) developmental arrest involving maternal effect mutations or hybrid incompatibility, (ii) the failure of parthenogenetic embryos, and (iii) the developmental potential of transferred nuclei from later embryos or adults, i.e. cloning. We include a brief discussion of some of the results obtained in the first two areas as they have an important bearing on the prospects for cloning. They have revealed that not only is the relationship between nucleus and cytoplasm important, but that between maternal and paternal genomes is also critical for correct development. In particular there are modulatory effects on gene activity that may be determined at some stage during gametogenesis, but which do not come into operation or show any effect until much later.

It might be expected that most early acting maternal effect mutations are due to deficiencies in components made during oogenesis. This was clearly shown to be the case for the 'o' mutation in the axolotl, where transfer of cytoplasm from a wild type oocyte can rescue the mutant embryos (Briggs and Cassens 1966; Brothers 1976). One maternal effect in the mouse that has now been analyzed by nuclear transfer is the incompatibility between eggs from the DDK strain with sperm from other strains (Wakasugi and Morita 1977). Very few embryos from such matings reach implantation, although if they do implant they develop normally. Reciprocal crosses or within DDK strain matings show normal litter sizes. When pronuclei from DDK ♀ × (CBA/C57)F_1♂ embryos were transferred to enucleated zygotes of non-DDK origin (F_1 × F_1) the majority of embryos implanted and developed normally (Mann 1985), This suggests that it is the DDK egg cytoplasm, rather than any maternal pronuclear component, that is involved in the incompatibility with foreign sperm.

Another maternal effect, but a much later acting one, is that of the T^{hp} mutation. When the mutant gene is inherited from the male parent most embryos survive, however, when it is·inherited via the female it is lethal, death occurring at late foetal stages or shortly after birth. McGrath and

Solter (1984a) demonstrated that this maternally inherited lethal effect of T^{hp} persists when T^{hp}/ + pronuclei are transplanted into + / + cytoplasm. If the defect is determined by the pronuclei and not the cytoplasm as this suggests, it implies that differential activation of the *T^{hp}* gene, depending on whether it underwent oogenesis or spermatogenesis, is sufficiently long lasting to produce an effect in later foetal life. This phenomenon of 'imprinting' was known for the X chromosome in female embryos where the paternally inherited X is preferentially inactivated in trophectoderm and extra-embryonic endoderm, as opposed to the random inactivation in the rest of the embryo (Takagi and Sasaki 1975; Harper *et al.* 1982).

Imprinting has also been demonstrated for a number of other chromosomes which may show lethal effects when both are inherited from one parent. Recently, Cattanach and Kirk (1985) have shown that mice with maternal duplication/paternal deficiency and its reciprocal for each of two particular chromosome regions show anomalous phenotypes which depart from normal in opposite directions. Thus, if two chromosomes 11 are of paternal origin the resulting newborn are considerably larger than normal, and if maternal they are much smaller. This effect has been localized to the proximal region of chromosome 11. The distal region of chromosome 2 shows a similar effect; if paternal the mice are squat and hyperkinetic, if maternal they are elongated and hypokinetic. Not all chromosomes show evidence of imprinting and it is clear in some cases that the activation of particular genes, e.g. glucose phosphate isomerase, is independent of the contributing parent (Gilbert and Solter 1985). These effects, however, are likely to have a bearing on our ability to carry out particular manipulations, such as deriving uniparental offspring or cloning.

The problem of the failure of parthenogenetic embryos has long perplexed researchers. Haploid or diploid parthenogenones, resulting from spontaneous or experimental activation of unfertilized eggs, will undergo apparently normal pre-implantation development, but die in the early post-implantation stages (Kaufman 1983). The furthest stage of development that has been reported is the 25 somite, forelimb bud stage (Kaufman 1981). However, in aggregation chimaeras with fertilized embryos, parthenogenetic embryos have the ability to differentiate into many tissue types, including functional gametes (Stevens 1978). One possible reason for the failure of parthenogenetic embryos was that sperm have some additional role, apart from restoring a complete genome, and so pronuclear transplantation was used to investigate whether sperm-related modifications to the egg cytoplasm are important. In these experiments, embryos produced by the transfer of pronuclei from diploid parthenogenetic eggs to enucleated fertilized eggs died soon after implantation; whereas viable young were obtained from the transfer of fertilized egg pronuclei into enucleated parthenogenetic eggs (Mann and Lovell-Badge 1984). This shows that the death of parthenogenones is not due to a lack of cytoplasmic factors from the sperm, but to the presence of two maternally-derived pronuclei. Diploid gynogenetic embryos, produced after the experimental removal of the paternal pronucleus, are equivalent to diploid parthenogenones in this respect, and show the same potential for development (Surani and Barton 1983; Surani *et al.* 1984). Embryos containing only two paternally-derived pronuclei also fail to develop, and it is the combination of maternal and paternal pronuclei which is necessary for normal development (McGrath and Solter 1984b; Barton *et al.* 1984; R. J. Mann, unpublished results).

It would now seem likely that it is the phenomenon of imprinting of the maternal and paternal genomes that is responsible for the failure of diploid parthenogenones. For the majority which die at implantation the cause cannot be prescribed to any particular chromosome. Under ideal conditions, or when delayed implantation or aggregation is used to increase cell number, it is possible to have up to 25 per cent beginning normal post-implantation development. We now have some evidence (Mann *et al.* 1985) that these eventually die because of a failure correctly to undergo X-inactivation. This may be attributed to the lack of a paternal X chromosome which in normal fertilized embryos is preferentially inactivated in extra-embryonic tissues. One might also expect that any parthenogenetic embryo that overcame this problem would then succumb to a combination of the phenotypes reported by Cattanach and Kirk (1985), probably at or just before birth.

The application of nuclear transfer technology that could have the greatest impact, both scientifically and economically, would be that of cloning. It may be used to establish the totipotency of nuclei, as has been done with amphibia, and provide information on the timing and processes of commitment. It could also be used to create a number of genetically identical individuals for studies where variation due to genetic background needs to be controlled, or for rapid expansion of a population with a particular desirable trait. This latter could be for reasons of scientific study, or where one has a novel breed of farm animal (transgenics, etc.) or a 'prizewinner'.

Cloning is very different from twinning or embryo splitting because it requires a change of developmental fate, or reprogramming. Whereas only a few identical individuals can be obtained from splitting techniques (Willadsen 1981), cloning would allow, if it can be performed sequentially, a large number to be produced. It is essential, however, to find both a suitable recipient cytoplasm that is able to reprogramme and donor nuclei that have not lost the ability to give rise to all cell types. As for the donor nuclei, they are most likely to work if they are taken from cells of the germ cell lineage. These cannot have undergone any permanent changes in DNA sequence, as would have occurred in a lymphocyte, for example, and they are by definition totipotent. Thus, one could consider cells of the early embryo, primordial germ cells, and premeiotic stages of gametogenesis. However, we now suspect that the nucleus has to be from a cell before 'imprinting' otherwise it might suffer the same fate as a uniparental embryo. Imprinting would logically occur during gametogenesis, but could be pre- or post-meiotic.

The first restriction in developmental potential of cells in the mouse embryo occurs between the eight-cell stage and the formation of the blastocyst with the differentiation of inner cell mass (ICM) and trophectoderm cells. Experiments involving the introduction of nuclei from these cells into non-enucleated one-cell fertilized embryos had suggested that those of the trophectoderm were no longer able to support development, although those of the ICM may be able to (Modlinski 1981). Indeed, it was claimed that development to term could be achieved after transfer of an ICM cell nucleus into an enucleated fertilized egg (Illmensee and Hoppe 1981). However, it has not been possible to repeat these experiments which were done with a technique that involved puncture of the egg plasma membrane, and it has been suggested that enucleation was incomplete (McGrath and Solter 1984c).

Using their technique, McGrath and Solter (1984c) have analyzed the potential of nuclei from cleavage stages and of inner cell mass cells after transfer into enucleated fertilized one-cell embryos. Whereas over 90 per cent of donor nuclei from one-cell embryos would allow development to the blastocyst stage *in vitro*, they found that this ability was rapidly lost during cleavage. Thus, only 19 per cent of two-cell nuclei supported development to the morula or blastocyst stage, and four-, eight-, and ICM-cell nuclei rarely gave development beyond the two-cell stage. They concluded that there was a loss of potential in the nuclei and, if true, this would imply that cloning was impossible. However, it is known that individual blastomeres of embryos up to at least the eight-cell stage are still totipotent, at least in chimaeras (Kelly 1975). It therefore seems more likely that the recipient cytoplasm is at fault and is unable to correctly reprogramme the transferred nuclei.

The one-cell embryo is peculiar in many respects. It has two transcriptionally inactive pronuclei, a long cell cycle time and specific translational controls which contribute to a unique pattern of protein synthesis (Bolton *et al.* 1984), so perhaps one should look at alternative recipients. It is possible to obtain good development to term after transferring nuclei of single blastomeres between two-cell embryos (R. J. Mann, unpublished observations), and some recent experiments suggest that enucleated two-cell embryos may allow development, at least to blastocysts, when later stage nuclei are used (Robl and First 1985). Alternatively, one could try an earlier stage, e.g. oocytes. This approach has worked well for cloning experiments in amphibia when the potential of adult cell nuclei has been tested (Di Berardino and Hofner 1983; Leonard *et al.* 1982). Development may even be improved when a second round of transfer back into oocyte cytoplasm is carried out (Di Berardino *et al.*, 1984). One simple hypothesis that could account for the mouse results is that maternal factors necessary for development are free in the cytoplasm of oocytes but they become sequestered into the pronuclei during their formation after fertilization (see De Leon *et al.* 1983). These factors will not be synthesized by the embryo, so later nuclei will contain insufficient amounts to allow development on transfer into the fertilized egg cytoplasm.

In conclusion, although attempts to clone mice have so far met with little success, there is still a lot of scope for experimentation and we should not be too pessimistic. Also, other species may be more amenable to this type of study. For example, while separate blastomeres from even four-cell mouse embryos are unable to give viable offspring (Kelly 1975) it is possible to obtain a newborn animal from an isolated blastomere of an eight-cell sheep embryo (Willadsen 1981). This reflects an improved capacity for size regulation and may mean that reprogramming would not be so essential for nuclei from such embryos.

Note added in proof: Willadsen has recently reported successful 'nuclear transplantation' in sheep embryos, using 8- and 16-cell blastomeres as donors into 'enucleated' halves of unfertilized oocytes, suggesting that cloning may indeed be possible. (Willadsen, S.M. (1986)). Nuclear transplantation in sheep embryos. *Nature* **320**, 63–5.

References

Barton, S.C., Surani, M.A.H., and Norris, M.L. (1984). Role of paternal and maternal genomes in mouse development. *Nature* **311**, 374–6.

Bolton, V.N., Oades, P.J., and Johnson, M.H. (1984). The relationship between cleavage, DNA replication and gene expression in the mouse 2-cell embryo. *J. Embryol. exp. Morph.* **79**, 139–63.

Briggs, R. and Cassens, G. 1966. Accumulation in the oocyte nucleus of a gene product essential for embryonic development beyond gastrulation. *Proc. Nat. Acad. Sci. USA* **55**, 1103–9.

—— and King, T.J. (1952). Transplantation of living nuclei from blastula cells into enucleated frogs' eggs. *Proc. Nat. Acad. Sci. USA* **38**, 445–63.

Brothers, A.J. (1976). Stable nuclear activation dependent on a protein synthesized during oogenesis. *Nature* **260**, 112–5.

Cattanach, B.M. and Kirk, M. (1985). Differential activity of maternally and paternally derived chromosome regions in mice. *Nature* **315**, 496–8.

Di Berardino, M.A. and Hoffner, N.J. (1983). Gene reactivation in erythrocytes: nuclear transplantation in oocytes and eggs of *Rana*. *Science* **219**, 862–4.

——, ——, and Etkin, L.D. (1984). Activation of dormant genes in specialized cells. *Science* **224**, 946–52.

De Leon, D.V., Cox, K.H., Angerer, L.M., and Angerer, R.C. (1983). Most early-variant histone mRNA is contained in the pronucleus of sea urchin eggs. *Dev. Biol.* **100**, 197–206.

Gilbert, S. F. and Solter, D. (1985). Onset of paternal and maternal Gpi-1 expression in preimplantation mouse embryos. *Dev. Biol.* **109**, 515–7.

Gurdon, J. B. (1962a). Adult frogs from the nuclei of single somatic cells. *Dev. Biol.* **4**, 256–73.

—— (1962b). The developmental capacity of nuclei taken from intestinal epithelium cells of feeding tadpoles. *J. Embryol. exp. Morph.* **10**, 622–40.

Hammer, R. E., Pursel, V. G., Rexroad Jr, C. E., Wall, R. J., Bolt, D. J., Ebert, K. M., Palmiter, R. D., and Brunster, R. L. (1985). Production of transgenic rabbits, sheep and pigs by microinjection. *Nature* **315**, 680–3.

Harper, M. I., Fosten, M., and Monk, M. (1982). Preferential paternal X inactivation in extra-embryonic tissues of early mouse embryos. *J. Embryol. exp. Morph.* **67**, 127–35.

Illmensee, K. and Hoppe, P. C. (1981). Nuclear transplantation in *Mus musculus*: developmental potential of nuclei from preimplantation embryos. *Cell* **23**, 9–18.

Kaufman, M. H. (1981). Parthenogenesis: a system facilitating understanding of factors that influence early mammalian development. In *Progress in anatomy*, Vol. 1 (eds. R. J. Harrison and R. L. Holmes) pp. 1–34. Cambridge University Press.

—— (1983) *Early mammalian development: parthenogenetic studies* Developmental and Cell Biology Series. 14, pp. 111–38. Cambridge University Press.

Kelly, S. J. (1975). Studies of the potency of early cleavage blastomeres of the mouse. In *The early development of mammals*, 2nd Symp. Br. Soc. Dev. Biol. (eds. M. Balls and A. E. Wild) pp. 97–106. Cambridge University Press, London.

Leonard, R. A., Hoffner, N. J., and Di Berardino, M. A. (1982). Induction of DNA synthesis in amphibian erythroid nuclei in *Rana* eggs following conditioning in meiotic oocytes. *Dev. Biol.* **92**, 343–55.

Mann, J. R. (1986). The DDK egg-foreign sperm incompatibility in mice is not between the pronuclei. *J. Reprod. Fert.* **76** (In press).

——, Burgoyne, P. S., and Lovell-Badge, R. H. (1986). The development of XO parthenogenetic and gynogenetic embryos constructed by nuclear transplantation. Submitted to *J. Embryol. exp. Morph.*

—— and Lovell-Badge, R. H. (1984). Inviability of parthenogenones is determined by pronuclei, not egg cytoplasm. *Nature* **310**, 66–87.

McGrath, J. and Solter, D. (1983). Nuclear transplantation in the mouse embryo by microsurgery and cell fusion. *Science* **220**, 1300–2.

—— and —— (1984a). Maternal Thp lethality in the mouse is a nuclear, not cytoplasmic, defect. *Nature* **308**, 550–2.

—— and —— (1984b). Completion of mouse embryogenesis requires both the maternal and paternal genomes. *Cell* **37**, 179–83.

—— and —— (1984c). Inability of mouse blastomere nuclei transferred to enucleated zygotes to support development *in vitro*. *Science* **226**, 1317–9.

Modlinski, J. A. (1981). The fate of inner cell mass and trophectoderm nuclei transplanted to fertilized mouse eggs. *Nature* **292**, 342–4.

Robl, J. M. and First, N. L. (1985). Transfer of nuclei from 8-cell to 2-cell murine embryos. *Theriogenology* **23**, 219.

Stevens, L. C. (1978). Totipotent cells of parthenogenetic origin in a chimaeric mouse. *Nature* **276**, 266–7.

Surani, M. A. H. and Barton, S. C. (1983). Development of gynogenetic eggs in the mouse: implications for parthenogenetic embryos. *Science* **222**, 1034–6.

——, ——, and Norris, M. L. (1984). Development of reconstituted mouse eggs suggests imprinting of the genome during gametogenesis. *Nature* **308**, 548–50.

Takagi, N. and Sasaki, M. (1975). Preferential inactivation of the paternally derived X chromosome in the extra embryonic membranes of the mouse. *Nature* **256**, 640–2.

Wakasugi, N. and Morita, M. 1977. Studies on the development of F$_1$ embryos from inter-strain crosses involving DDK mice. *J. Embryol. exp. Morph.* **38**, 211–6.

Wall, R. J., Pursel, V. G., Hammer, R. E., and Brinster, R. L. (1985). Development of porcine ova that were centrifuged to permit visualization of pronuclei and nuclei. *Biol. Reprod.* **32**, 645–51.

Willadsen, S. M. (1981). The developmental capacity of blastomeres from 4- and 8-cell sheep embryos. *J. Embryol. exp. Morph.* **65**, 165–72.

Zimmermann, H. and Vienken, J. (1982). Electric field-induced cell-to-cell fusion. *J. Memb. Biol.* **67**, 165–82.

Discussion

Brem asked about the fate of maternal RNA in early embryos. Lovell-Badge replied that it is degraded rapidly in early development and most is thought to have gone by the morula stage.

Hanset asked about the incidence of viable parthenogenotes in other species, especially birds such as turkeys? Lovell-Badge said that while it is possible to have parthenogenetic turkeys, they are not very fit. We need to know more about the very early development of birds, state of sex-chromosome activation, etc., before we can relate this to the situation in the mouse. Production of parthenogenotes is certainly much easier in Amphibia. Perhaps the imprinting of paternal and maternal chromosomes as described by Cattanach is important only for extra embryonic membranes.

Land asked how serious the technical problems in establishing cloning in mice are, how are they to be overcome and whether there are further problems in extending the technique to farm animals? Lovell-Badge was not so pessimistic about overcoming the technical problems in mice. Although there is doubt surrounding it, some work by Illmensee has suggested possible ways of overcoming technical barriers to cloning, e.g. by disrupting the two pronuclei. We should also try the other stages as recipients. In general, a lot more work needs to be done. The studies by Willadsen have suggested that, in some respects, the process may be simpler in sheep. There is a general shortage of information in farm species.

Smith asked how important cytoplasmic effects are, but it seems little is known apart from the case of incompatibility in the DDK strain of mice. Hill wondered if cloning from cultured cells, such as EK cells (embryonic stem cells), offered more or less prospect of success. Lovell-Badge thought it worth trying. They can be used to give chimaeras which are comprised of genetically different hosts and a proportion of genetically identical EK-derived cells. A number of groups are attempting to derive EK typed cells in farm animals. There has been no progress in cloning cells from mature animals rather than embryonic cells. Land wondered about the time scale in application of techniques in an animal breeding programme. It was felt that this was hard to assess, but that progress was fast and would be faster as more groups work on farm species.

12

Transposable elements in genetic selection

Trudy F. C. Mackay

Abstract

The P family of transposable elements in Drosophila melanogaster *transpose with exceptionally high frequency when males from P strains carrying multiple copies of these elements are crossed to females from M strains which lack P elements, but with substantially lower frequency in the reciprocal crosses. Transposition is associated with enhanced mutation rates, and may create new variation for quantitative traits; if so, accelerated response to artificial selection of progeny of dysgenic (M♀♀ × P♂♂) crosses is expected compared to that from progeny of non-dysgenic (P♀♀ × M♂♂) crosses. Divergent artificial selection for number of bristles on the last abdominal tergite was carried out for 16 generations among the progeny of dysgenic and non-dysgenic hybrids, each cross replicated four times. Average realized heritability of abdominal bristle score for the crosses in which P transposition was expected was 0.244 ± 0.017, 1.5 times greater than average heritability estimated from crosses in which P transposition was expected to be rare (0.163 ± 0.010). Much of the new mutational variation was associated with deleterious effects on fitness. Several putative 'quantitative' mutations were identified from chromosomes extracted from the selected lines, which will form the basis for further investigation at the molecular level of the genes controlling quantitative inheritance.*

Introduction

One of the more intriguing results of molecular analysis of eukaryotic DNA is the discovery that many sequences are apparently transposable. (For a comprehensive review of mobile sequences, see Shapiro 1983.) Transposable sequences, or elements, are typically present as moderately repeated (from 10 to 100 copies) sequences scattered at random throughout the genome. At least five major classes of transposable element have been identified in *Drosophila*, and together they comprise approximately one-fifth of the genome (Rubin 1983). These elements have been labelled 'parasitic' or 'selfish' DNA (Doolittle and Sapienza 1980), since replicative transposition ensures their own survival, but they have no discernable phenotypic effect on their host until they move. Movement of sequences may cause deleterious mutations by altering host DNA, since insertion of an element may disrupt a structural locus or alter its regulation, and imprecise excision often generates deletions and chromosomal rearrangements (Rubin 1983). To what extent does movement of these elements affect the loci controlling quantitative traits?

It is possible to determine the effect of mobilization of transposable elements on polygenic loci in *Drosophila* by studying the results of P element movement. The P family of transposable elements in *Drosophila melanogaster* consists of a large 2.9 kb sequence, and various smaller sequences, each of which can be derived by different internal deletions of the large element (O'Hare and Rubin 1983). P elements are unusual among transposable elements so far discovered in *Drosophila* in that they are not present in all strains. Strains of *D. melanogaster* which contain these elements ('P' strains) typically have 30–50 copies per haploid genome (the exact number and position varying across strains), while other strains ('M' strains) have no functional P elements (Bingham *et al.* 1982). When P males are crossed to M females the F_1 progeny are characterized by a number of abnormalities which have been embraced by the term 'hybrid dysgenesis' (Kidwell *et al.* 1977), and which are associated with high rates of transposition of P (and perhaps other) elements. Rates of P transposition are not enhanced in the F_1 hybrids of the reciprocal cross of M males to P females, or in intra-P (and of course intra-M) crosses. If movement of P elements causes mutations affecting polygenic loci controlling a particular quantitative trait, dysgenic (M ♀ ♀ × P ♂ ♂) and non-dysgenic (P ♀ ♀ × M ♂ ♂) hybrids should differ genetically mainly because of P-induced mutations in the former cross. Therefore, comparison of response to artificial selection for that trait in dysgenic and non-dysgenic hybrids should lead to the estimation of the amount and nature of P element-induced quantitative variation. The major features of the first 16 generations of response to divergent artificial selection for abdominal bristle number among the progeny of dysgenic and non-dysgenic hybrids are reviewed here; for a more detailed summary see Mackay (1985).

Materials and methods

Harwich (P) and Canton S (M) strains of *D. melanogaster* were used to set up two replicates of non-dysgenic (10 P ♀ ♀ × 10 M ♂ ♂) and dysgenic (10 M ♀ ♀ × 10 P ♂ ♂) crosses. The following generation (GO), 50 individuals of each sex were scored for abdominal bristle count on the last abdominal tergite, and the ten highest scoring males and females, and ten lowest scoring males and females, were crossed *en masse* to found 'non-dysgenic' and 'dysgenic' high and low selection lines, one pair of lines for each replicate. Selection was continued in subsequent generations by choosing the ten most extreme individuals of 50 scored of each sex, in each line, to be parents of the next generation. An additional two replicates were formed by initially crossing Harwich (P) and Canton S (M) sublines that had been inbred by full-sibbing for eight generations, otherwise the selection procedure was as described above. The first two replicates are subsequently referred to as the 'non-inbred' crosses, while the second pair of replicates are the 'inbred' crosses. Selection was continued in all lines for 16 generations. The flies were reared on cornmeal-agar-molasses medium, and all cultures were incubated at 20°C, a temperature at which gonadal sterility is not appreciable in dysgenic hybrids.

Data was collected from generations 10–12 of the 'non-inbred' selection lines on response of the lines to relaxed selection, and homozygous viability and bristle score of second and third chromosomes extracted from the selection lines. For details of the procedures see Mackay (1985).

Results

Generation means of the 'non-inbred' and 'inbred', non-dysgenic and dysgenic selection lines are shown in Figs 12.1 and 12.2. The response to selection for each of the dysgenic replicates is on average twice that of the

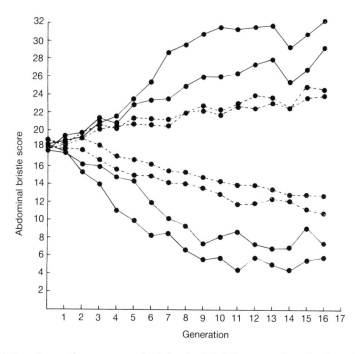

Fig. 12.1. Generation means of abdominal bristle score for the 'non-inbred' dysgenic (solid lines) and non-dysgenic (broken lines) selection lines.

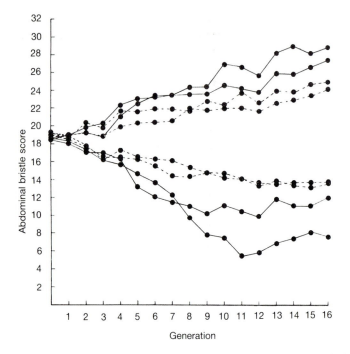

Fig. 12.2. Generation means of abdominal bristle score for the 'inbred' dysgenic (solid lines) and non-dysgenic (broken lines) selection lines.

non-dysgenic replicates, and the pattern of the response differs. Variation in response among dysgenic replicates is greater than that among non-dysgenic replicates, and most of the response of the dysgenic crosses occurs in the first ten generations of selection in both directions, compared to the non-dysgenic crosses in which the response is more nearly linear over the 16 generations.

Realized heritabilities of the dysgenic and non-dysgenic crosses, calculated from regression of cumulated divergence in response between high and low lines on cumulated selection differential (Falconer 1981) are compared, therefore, based on response to generation 10 (Table 12.1). Realized heritability is on average 1.5 times greater for the dysgenic selection lines.

Table 12.1. Realized heritabilities of abdominal bristle score, calculated as described in the text

	Replicate	Non-dysgenic lines	Dysgenic lines
'Non-inbred' lines	1	0.167	0.280
	2	0.190	0.214
'Inbred' lines	1	0.152	0.216
	2	0.142	0.264

The apparent plateaux in response of the dysgenic selection lines after generation 10 are not associated with a reduction in phenotypic variance. Phenotypic variance of abdominal bristle number increases in the dysgenic selection lines over the first six generations of selection to a level two to five times that of the non-dysgenic lines, and remains so despite continued selection.

Much of the P element-induced variation in the dysgenic lines is associated with deleterious effects on fitness. Relaxation of artificial selection at generation 10 for six generations had no effect on average bristle score of the non-dysgenic lines, but mean performance of the dysgenic high lines declined over this period to the level of the non-dysgenic high lines, and mean score of the dysgenic low lines rose appreciably, although remaining lower than the non-dysgenic populations. This immediate response to natural selection is consistent with the elimination of deleterious alleles which nevertheless have large effects on the character. Further evidence for reduced fitness of the dysgenic, compared to non-dysgenic lines, comes from estimates of second and third chromosome homozygous viabilities. Twice as many chromosomes extracted from the dysgenic lines were lethal when homozygous, and viabilities of non-lethal dysgenic chromosomes were significantly reduced compared to those isolated from non-dysgenic lines (Mackay, 1985).

Several chromosomes with extreme abdominal bristle scores were extracted from high and low dysgenic selection lines, but not the non-dysgenic lines. The most extreme effects were detected on five female-sterile, poorly viable chromosomes isolated from the dysgenic low lines, with an average score of two abdominal bristles (on all tergites). The allele on these chromosomes is allelic to the smooth locus (Lindsley and Grell 1968; Alan Robertson, personal communication). The difference in response of the dysgenic and non-dysgenic crosses to selection for decreased abdominal bristle number cannot, however, be explained entirely in terms of the segregation of alleles at the smooth locus in the dysgenic lines, since the recessive, semi-lethal smooth allele cannot be maintained as a homozygote. Furthermore, a population segregating for only wild-type and mutant smooth alleles would have a bimodal distribution of bristle phenotypes, not the nearly uniform distribution observed.

Discussion

It appears that a significant amount of new mutational variation affecting the quantitative trait, abdominal bristle number, can be generated by the activity of transposable elements mobilized during hybrid dysgenesis. Although much of this new variation was deleterious, sufficient new additive variation was generated to account for an accelerated response to artificial selection in dysgenic, compared to non-dysgenic hybrids. How does the amount of variation arising from dysgenesis compare with that created by more conventional mutagens? Clayton and Robertson (1955, 1964), and Hollingdale and Barker (1971) have investigated the effect of X-irradiation on the induction of selectable variation for abdominal bristle number of *D. melanogaster*. Comparison of response to divergent artificial selection of control and irradiated (Clayton and Robertson, 1800 r/generation; Hollingdale and Barker, 1000 r/generation) populations consistently gave estimates of an increase in response of the treated populations, over approximately 20 generations of irradiation, of about 1.5 bristles (single segment). This corresponds to an increment in additive genetic variation affecting single segment bristle score of 2.4×10^{-6}/r/generation (Clayton and Robertson 1964). The increase in response of the dysgenic populations averages 10.8 bristles (divergence) greater than the mean response of the non-dysgenic lines in only 10 generations, and a conservative estimate of the new additive variance arising from activities of transposable elements is 1.8 (Mackay 1985). It would therefore take a cumulative dose of 750 000 r of X-rays to produce an equivalent amount of new variation to that caused by dysgenesis. It is doubtful that the animals would survive such massive doses of X-rays, yet they have stood up to the element-induced mutagenesis. Perhaps the elements are truly selfish.

What is the nature of this new variation? Dysgenesis causes mutations by insertion and deletion of P elements, as well as chromosome rearrangements, all of which could potentially contribute to variation for abdominal bristle number. The absolute number of mutational events may be very large. Transposition is likely to occur for several generations after an initial dysgenic cross, as it may take up to ten generations to establish the P cytotype necessary for relative stability of the elements (Kidwell *et al.* 1981; Kiyasu and Kidwell 1984). There is also evidence, as yet circumstantial, that P element transposition mobilizes other families of transposable elements in concert (Rubin *et al.* 1982; Gerasimova *et al.* 1984).

Because substantial transposable element-induced variation has been detected for the quantitative character, abdominal bristle number, further analysis of the selection lines may lead to the identification and molecular characterization of genes controlling quantitative inheritance. Bingham *et al.* (1981) have proposed a novel method for studying any genetically defined locus at the molecular level, providing that a mutation of the gene of interest is caused by insertion of a transposable element, which itself has been cloned. The inserted element plus flanking sequences corresponding to the wild-type can then be recovered from a clone library by homology to the transposable element sequence. Element sequences can be subsequently removed from this clone by restriction endonuclease digestion to create a probe containing the fragment of wild-type sequence, which can then be used to recover the unmutated gene of interest. This strategy has been applied successfully to the cloning and sequencing of the white locus of *D. melanogaster* (Bingham *et al.* 1981), but is not restricted to Mendelian phenotypes and could in principle be extended to mutant alleles at loci controlling quantitative traits.

How successful is P element mutagenesis likely to be as a method for recovery of insertional mutations at loci controlling quantitative traits? This depends on several factors. Firstly, it is necessary that the relevant mutations are caused by insertion of elements, and not deletion or rearrangements. Secondly, success of the scheme depends on the sites of insertion of the elements being scattered at random throughout the genome, so many loci are likely to be affected. Finally, the insertion must be recognizable. One way to identify quantitative mutations is to extract chromosomes from the dysgenic and non-dysgenic selection lines and score their homozygous and heterozygous phenotypes. Any chromosome isolated from a dysgenic line with a phenotypic effect larger than those observed for chromosomes of non-dysgenic lines is a candidate for further study, but insertions causing small phenotypic changes may be missed. The smooth mutation is an example of a quantitative mutation identified in this way; one hopes it proves to be caused by an insertion. The second way to identify putative insertions affecting polygenes is to compare locations of P elements in the parental (Harwich) strain with their sites in the dysgenic and non-dysgenic high and low selection lines, by *in situ* hybridization of P DNA to polytene salivary gland chromosomes. By this method one would hope to identify any new sites of insertion which on further analysis may prove to be at loci of interest.

Although there are problems associated with the interpretation of difference in response to selection of dysgenic and non-dysgenic hybrids as due mainly to mobilization of elements in the former populations (Mackay 1985), the general approach combining quantitative genetic and molecular methodologies may in the future yield some insight into the molecular basis of quantitative inheritance.

References

Bingham, P.M., Kidwell, M.G., and Rubin, G.M. (1982). The molecular basis of P-M hybrid dysgenesis: the role of the P element, a P-strain-specific transposon family. *Cell* **29**, 995–1004.

——, Levis, R., and Rubin, G.M. (1981). Cloning of DNA sequences from the *white* locus of *D. melanogaster* by a novel and general method. *Cell* **25**, 693–704.

Clayton, G.A. and Robertson, A. (1955). Mutation and quantitative variation. *Am. Nat.* **89**, 151–8.

—— and —— (1964). The effects of X-rays on quantitative characters. *Genet. Res.* **5**, 410–22.

Doolittle, W.F. and Sapienza, C. (1980). Selfish genes, the phenotype paradigm, and genome evolution. *Nature* **284**, 601–3.

Falconer, D.S. (1981). *Introduction to Quantitative Genetics* (2nd edn). Longman, London.

Gerasimova, T.I., Mizrokhi, L.J., and Georgiev, G.P. (1984). Transposition bursts in genetically unstable *Drosophila melanogaster*. *Nature* **309**, 714-6.

Hollingdale, B. and Barker, J.S.F. (1971). Selection for increased abdominal bristle numbers in *Drosophila melanogaster* with concurrent irradiation. Theoret. Appl. Genet. **41**, 208-15.

Kidwell, M.G., Kidwell, J.F., and Sved, J.A. (1977). Hybrid dysgenesis in *Drosophila melanogaster*: a syndrome of aberrant traits including mutation, sterility, and male recombination. *Genetics* **86**, 813-33.

——, Novy, J.B., and Feeley, S.M. (1981). Rapid unidirectional change of hybrid dysgenesis potential in *Drosophila*. *J. Hered.* **72**, 32-38.

Kiyasu, P.K. and Kidwell, M.G. (1984). Hybrid dysgenesis in *Drosophila melanogaster*: the evolution of mixed P and M populations maintained at high temperature. *Genet. Res.* **44**, 251-9.

Lindsley, D. L. and Grell, E.H. (1968). *Genetic variations of Drosophila melanogaster*. Carnegie Inst. Wash. Publ. 627.

Mackay, T.F.C. (1985). Transposable element-induced response to artificial selection in *Drosophila melanogaster*. *Genetics* **111**, 351-374.

O'Hare, K. and Rubin, G.M. (1983). Structures of P transposable elements and their sites of insertion and excision in the *Drosophila melanogaster* genome. *Cell* **34**, 25-35.

Rubin, G.M. (1983). Dispersed repetitive DNAs in *Drosophila*. In *Mobile Genetic Elements* (ed. J.A. Shapiro). Academic Press, New York. pp. 329-361.

——, Kidwell, M.G., and Bingham, P.M. (1982). The molecular basis of P-M hybrid dysgenesis: the nature of induced mutations. *Cell* **29**, 987-94.

Shapiro, J.A. (1983). *Mobile Genetic Elements*. Academic Press, New York.

Discussion

Di Berardino wondered if transposable elements occur in mammals, but they do not. However, retroviruses may perform a similar role and these occur in all mammalian species. Stam asked what the copy number of P elements is in the selected lines. Preliminary data indicate that generally there are more than 10 elements and probably 30-50 copies as in the parental P strain. King asked if the dysgenic and non-dysgenic lines differ in fitness. They do, with the dysgenic being much poorer.

Stam also wondered how in the presence of so many deleterious mutations, was such a good response produced. MacKay replied that the deleterious effects on fitness may be due to lethals or other mutations at loci linked to genes affecting bristle number. Smith wondered if selection were relaxed later, whether the loss of response would be less. However, this could not be tested since the selection was stopped at generation 16. Linked fitness effects might be lost by recombination during selection but pleiotropic effects on fitness at bristle loci will persist.

Other experiments with P elements have shown that their insertions have often dramatic negative effects on fitness. Robertson suggested that the same is true for transgenic mice where the inserted genes may have deleterious effects on fitness and may be difficult to make homozygous. Lovell-Badge agreed and said that about one-third of transgenic insertions have a measurable effect on fitness. Smith wondered if P elements had contributed to responses in other selection experiments. MacKay thought probably, those experiments using crosses between different strains as the base population being most suspect. For instance, Scosserelli found a difference in response of reciprocal crosses where he used strains now known to be I and R types. (The I element is the other class of transposable element in *Drosophila melanogaster* which causes hybrid dysgenesis. R strains have no active I elements.)

13
Genetic engineering applied to milk producing animals: some expectations

J.C. Mercier

Abstract

In recent years, new techniques for engineering and introducing functional foreign genes into whole organisms have been developed and used successfully for producing transgenic mice. This new technology raises animal breeders' expectations for the rapid improvement of farm animals. Some potential applications of the new technology for improving the quality of milk both for nutritional and for industrial purposes are briefly described. Special emphasis has been given to the lactose problem in order to illustrate the potential of the new techniques and the strategies that may be used to generate transgenic animals producing milk with a lower lactose content.

Introduction

Milk is the universal food of newborn mammals, and its composition differs qualitatively and quantitatively among species, which suggests some adaptation to the specific nutritional requirements of the offspring. Animal milk has been included in human diet since early history. Today milk and the products derived from it provide up to 30 per cent of dietary proteins in developed Western countries and are important worldwide in infant nutrition. To meet his needs, man has genetically improved some species for milk production. Conventional methods of selection have proved very effective in that the genetic ability of dairy cattle to produce large quantities of milk has been dramatically improved and marked differences in milk composition among breeds have been obtained. More recently artificial insemination and embryo transfer have been effectively employed in this regard.

Nevertheless, as long as the choice of animals to be parents of the next generation is restricted to the existing population, dairy cattle breeders cannot hope to change the composition of milk significantly in order to meet the fast-changing nutritional and industrial requirements. Such a goal requires the artificial modification of the dairy animal genotype, and the novel technologies of genetic engineering and gene transfer seem to be promising in this respect.

The application of recombinant DNA technology has allowed the identification, isolation, and characterization of specific genes. Currently, new systems of gene transfer which allow the insertion of functional foreign genes into the mammalian genome and their subsequent transmission through the germ line are being developed (Gordon and Ruddle 1985). This technology opens up new possibilities for the study of gene regulation and also raises animal breeders' expectations for the rapid genetic improvement of farm animals (Palmiter *et al.* 1982; Hammer *et al.* 1985).

The techniques for engineering and introducing foreign genes into embryo cells as well as their potential use for generating transgenic animals able to produce more milk or foreign proteins of biological interest have been described elsewhere (e.g. Lathe *et al.*, 1986. Chapter 10). The principal object of this article is to discuss some possible applications of this new technology for improving the quality of milk both for nutritional and for industrial purposes. Special emphasis will be given to the lactose problem in order to illustrate the potential of this approach. As background to the discussion, our current knowledge of the biochemistry of lactation will be summarized.

Limitations of conventional methods of breeding for milk production and for improvement of milk quality

Since progress from selection is directly dependent on the amount of variation in the population for a given trait, the magnitude of the heritability, the genetic correlations between the different traits, and the selection intensity that can be practiced, there are limitations inherent in the conventional methods of breeding. (1) The heritabilities of yields of milk, total solids, fat, protein and lactose are of average magnitude (20–30 per cent) whereas the heritabilities of the percentages of these components are approximately twice these values (Touchberry 1974). (2) These genetic traits are expressed in only one sex so the evaluation of the breeding value of the male requires several years (6–7). (3) These factors are controlled by many genes. Those for milk volume are not identical with those controlling the various aspects of milk composition so simultaneous selection for several characteristics slows progress unless the genetic correlation of the factors is high. Furthermore, the positive correlation between the percentages of fat and protein is now considered as a disadvantage because current nutritional and industrial standards require milk with a low fat and a high protein content. (4) Selective breeding is a slow and tedious process. Consequently, the animal and/or the product gradually devised by the breeders with regard to current economic criteria may not necessarily meet the future nutritional and industrial requirements of consumers and manufacturers. For example, in the past, most dairy cattle breeders selected simultaneously for milk production and butterfat content at a time when the economic value of milk depended largely on lipids, whereas today the protein content has gained much more importance because of increasing demand for cheese and the rejection of animal fats by many consumers.

Because of these limitations, selection alone would not bring about great changes in the composition of milk such as the relative ratios of protein to fat or lactose to water.

Overview of the present state of knowledge in the field of the biochemistry of lactation

General understanding of the hormonal control of lactogenesis, the regulation of expression of milk protein genes and the biosynthesis, and secretion of milk components, has advanced rapidly in recent years.

The lactating mammary gland secretes a large number of proteins. Some originate from the blood and others are synthesized in large amounts by mammary epithelial cells. In ruminant species, nearly 95 per cent of total milk protein output is comprised of six proteins. These are the four caseins, α_{s1}, α_{s2}, β and κ (encoded by four clustered autosomal genes), β-lactoglobulin and α-lactalbumin. These proteins undergo covalent modifications prior to their packaging into secretory vesicles. These modifications include the co-translational proteolytic processing of signal peptides (which are in a way the visas that allow their protein owners to pass through endoplasmic reticulum border only once); the co- and post-translational N-glycosylation of α-lactalbumin at the recognition sites -Asn-X-Thr/Ser-; the post-translational O-phosphorylation of -Ser/Thr-X-A- triplets (where A is an acidic residue) of the four caseins, and the O-glycosylation of κ-casein. In the lumen of Golgi cisternae, κ-casein prevents precipitation of the other caseins with calcium by forming co-polymers or submicelles which further aggregate into stable micelles. This occurs via presumed bridges between phosphoseryl residues and clusters of tricalcium phosphate $[Ca_3(PO_4)_2]_3$ (Schmidt 1982). Lactose is synthesized at the same site according to the reaction 'UDP-galactose + Glucose → Lactose + UDP'. This reaction is promoted by α-lactalbumin which modifies the substrate specificity of a membrane-bound galactosyltransferase that normally uses N-acetylglucosamine residues as galactose acceptors. The secretory vesicles containing casein micelles, other proteins, lactose, etc., ultimately fuse with the apical plasma membrane and release their contents into the acinar lumina which are linked to the duct system of the mammary gland.

The genes encoding the milk proteins contain many introns (three in rat and bovine α-lactalbumin genes; three in rat and mouse whey acidic protein genes; at least nine in rat γ-casein gene). Their expression is controlled by several hormones. In ruminants, prolactin induces transcription of milk protein genes, increases the stability of the relevant mRNAs, and favours specifically their translation. This induction is amplified by glucocorticoids, which are inactive alone, and is inhibited by progesterone.

For more detailed information, the reader is referred to recent books (Fox 1982; Mepham 1983) and citations therein.

Some important nutritional and technological problems involving milk components: prospects for genetic engineering

Reviewing all possible modifications of the dairy animal genome, and their presumed effects upon the nutritional and technological qualities of milk is beyond the scope of this paper. Only three important potential applications of the new techniques of genetic engineering and gene transfer will be outlined. (1) Reduction of the lactose content of milk to promote its consumption by people intolerant of lactose and to facilitate the manufacture of lactose-free products. (2) Suppression of β-lactoglobulin in bovine milk used for preparing humanized infant formulae. (3) Production of milk with high casein and organic phosphate contents.

Lowering the lactose content of milk

Nutritional and industrial implications of the high lactose content of milk
Lactose content of milk varies greatly among species (no lactose in California sea lion; 1–10 g/l in cetacea; 40–50 g/l in domestic ruminants; 70 g/l and over in many primates; 75 g/l in man), and there is a reciprocal relation between lactose and osmotic salt concentrations.

After milk ingestion, lactose is normally hydrolyzed into glucose and galactose by a transient dimeric enzyme (lactase of β-galactosidase) lining the luminal side of the jejunal brush-border membrane of the suckling young, and to a lesser extent by the microflora of the large intestine. Lactase monomers (mol. wt 160 000 for human) may be derived from a single primary polypeptide chain (mol. wt 245 000) by means of N-glycosylation and post-translational proteolytic processing (Skovbjerg *et al.* 1984). Their synthesis appears to be controlled by an autosomal temporal regulatory gene (Flatz 1983). In man, a widespread 'normal hypolactasia' allele ℓ may control the post-weaning switch from high to low lactase synthesis (up to 95 per cent repression) whereas a dominant allele L occurring in Indo-European populations may be responsible for persistent lactase production. Consequently, adult individuals with genotypes LL and Ll are still able to digest lactose whereas homozygotes ll prevalent in most African, Asian and American-Indian populations are lactose-intolerant. They suffer from severe digestive disturbances associated with osmotic pressure of non-digested lactose (bloating and diarrhoea) and increased metabolic activity of the normal intestinal flora (fermentations).

Post-weaning lactase repression and congenital alactasia (lack of functional lactase) aside, other forms of lactose-intolerance are pathological and result from alteration of the brush-border membrane in adults or infants suffering from gastroenteritis or malnutrition. According to recent estimations (Dahlqvist 1983), lactose malabsorbers may represent up to 90 per cent of the human race (3 per cent of Swedes and Danes, 20–30 per cent of Britons, 42 per cent of French, 6 per cent of U.S.A. whites, and 73 per cent of U.S.A. negroes, nearly 100 per cent of Japanese and people from the whole Far East).

The physicochemical properties and the high concentration of lactose in milk set limits to many industrial processes and contribute to the polluting strength of the whey. Manufacture of high-quality products for several industrial and food applications requires partial or complete removal of lactose (Muller 1982; Marshall 1982; Morr 1982), which is achieved through various costly procedures.

Lactose is a matter of great concern in the dairy industry as illustrated by the current extensive research carried out on lactose hydrolysis of milk and whey by means of free or immobilized β-galactosidases from a wide range of sources (Richmond *et al.* 1981; Gekas and Lopez-Leiva 1985). Tests with 'lactose-hydrolyzed milks', now marketed in some countries, have shown that they are well-tolerated by lactose-malabsorbers (reviewed in Delmont 1983). It has been observed that a high level of galactose in the diet induces cataract in rats, but such harmful effects do not occur in man when the diet contains both galactose and glucose (quoted in MacDonald and Williams 1983; Delmont *et al.* 1983), which is the case of 'lactose-hydrolyzed milk'.

The production of a raw milk with a reduced lactose content would be advantageous for both consumers and manufacturers on nutritional, technological, commercial, and environmental grounds, and could promote its consumption since milk rejection today is generally attributed to both socio-cultural factors and lactose intolerance.

Lowering the lactose content of milk by means of germline manipulation   In principle, a decrease in lactose synthesis might be obtained by decreasing the amount of α-lactalbumin available in the Golgi apparatus or by altering its ability to interact with the UDP-galactosyltransferase. This might be accomplished by structural modifications of the DNA sequences encoding respectively the promoter-enhancer region, the signal peptide, and the protein domain interacting with the enzyme. Unfortunately, in higher eukaryotes, it is not yet possible to replace a normal gene with a counterpart that has been altered *in vitro*. This problem, however, might be overcome by using two different approaches which require the construction of novel DNA molecules and their functional insertion into the dairy animal genome.

(1) Inhibition of α-lactalbumin gene expression by anti-sense RNA. Two complementary strands of RNA can associate to form a stable duplex. This is

one of the mechanisms whereby prokaryotes regulate gene expression as hybrid formation between mRNA and the anti-sense RNA prevents initiation of translation (reviewed by Travers 1984). Inhibition of gene expression by artificial anti-sense RNA has been successfully demonstrated in eukaryotes (Laporte 1984; Weintraub *et al.* 1985 and citations therein; Melton 1985; Rosenberg *et al.* 1985) and short lengths of anti-sense RNA complementary to the 5′ untranslated region of mRNA appear to be the most effective (Weintraub *et al.* 1985).

Preliminary experiments must be carried out to determine the most effective lengths of anti-sense transcript, its stability, and the ratio anti-sense RNA/α-lactalbumin mRNA necessary for complete inhibition. For this purpose, two types of plasmid vectors containing the SV40 early promoter and polyadenylation signal, a suitable intron, a dominant selectable marker, and α-lactalbumin cDNA cloned in both orientations, may be constructed and tested in co-transfected cell cultures (e.g. fibroblast cell lines). Ultimately, an anti-sense α-lactalbumin gene controlled by the strong promoter-enhancer sequences of a gene encoding a milk protein (α-lactalbumin, β-lactoglobulin, or a casein) and transcribed as anti-sense α-lactalbumin RNA may be constructed and micro-injected into the pronuclei of animal eggs to generate transgenic animals (Constantini and Lacy 1981; Palmiter *et al.* 1982; Hammer *et al.* 1985). Expression of both the exogenous gene and the set of normal endogenous genes encoding milk proteins would occur under the control of the same hormones, resulting in the concomitant transcription of both the α-lactalbumin gene and its anti-sense counterpart, and the inhibition of α-lactalbumin mRNA by duplex formation.

(2) Secretion of an active β-galactosidase into milk. Animals may be genetically engineered to synthesize and secrete an active enzyme able to hydrolyze lactose under physiological conditions. This approach requires in the first place the construction of a hybrid gene resulting from the in-frame insertion of the DNA sequence encoding the signal peptide of a milk protein (e.g. a casein or β-lactoglobulin) into a suitable β-galactosidase gene fused to the SV40 early promoter (see for example, Rubenstein *et al.* 1984). The construct may be tested in appropriate cell lines to determine whether an active enzyme is secreted. Ultimately, the hybrid gene may be fused to the promoter-enhancer region of a milk protein gene and the final construct micro-injected into the pronuclei of animal eggs to generate transgenic animals. In principle, milk of these animals would contain an active β-galactosidase secreted concomitantly with milk proteins, and able to convert lactose into glucose and galactose, thus reducing the lactose content of milk.

In this way, milk would contain a β-galactosidase introduced under safe conditions, which is not always the case for the various enzyme preparations used industrially.

Modifications of cow's milk to simulate some characteristics of human milk

The gradual decline of breast-feeding in human populations has resulted in an increased use of human milk substitutes. As mentioned earlier, composition of milk differs widely among species (e.g. the average ratios of some nutrients concentrations in cow's and human milks are respectively: proteins 3:1; caseins 7:1; whey proteins 1.2:1; lactose 1:1.6. In addition, human milk lacks α_{s1}-, α_{s2}-casein, and β-lactoglobulin, and has a high content of lactoferrin and lysozyme), so raw farm animal's milk is inadequate for the infant diet. Its ingestion by the human newborn results in intestinal disorders and possibly allergic responses.

Most current infant formulae, the so-called humanized milk formulae, are largely based on bovine milk that is altered to imitate human milk. Thus, dilution of cow's milk and addition of extra lactose, whey proteins, and possibly lactoferrin and lysozyme, make it more comparable to its human counterpart with respect to the casein/whey protein ratio. One potential disadvantage is the addition of β-lactoglobulin into the infant diet. This protein, which is a foreign substance for human, may be responsible for many observed allergies to cow's milk (Saperstein 1974; Hambraeus 1982; Moneret-Vautrin 1983) and is difficult to digest.

Elimination of β-lactoglobulin from whey would solve this problem, but the process is difficult to set up at the industrial level. This problem may be overcome by creating transgenic animals unable to synthesize β-lactoglobulin. The inhibition of β-lactoglobulin expression may be obtained by using the anti-sense RNA methodology outlined in the preceding section.

Other possible modifications of milk composition and milk proteins

Genetic engineering offers potential for many other types of manipulation dealing with the yield and/or the physico-chemical properties of milk proteins. It should be possible to introduce an extra casein gene into the chromosomes of a dairy animal and to increase the content of the relevant casein in milk and consequently the yield of curd available for cheese-making.

Moreover, the gene may be modified for improving the functional properties of the casein product. For example, the in-frame insertion of a few nucleotide sequences encoding the clustered phosphorylation recognition sites -Ser-Ser-Ser-Glu-Glu- (Mercier 1981) would result in an increase of the phosphate content of the relevant casein. It is likely that such a modification may contribute to improve both the stability of micelles and the calcium intake in the digestive tract.

Conclusion

In view of the preliminary results obtained in modifying the germline of a few animals, mice and more recently rabbits, pigs, and sheep, the novel technologies of genetic engineering and gene transfer appear to be of great promise for improving farm animals. However, model systems such as mammalian cell cultures and transgenic mice will have to be used to assess the feasibility of the strategies devised for engineering and introducing functional genes into whole organisms. Much progress remains to be made before these promising techniques are applied directly and routinely to generate transgenic farm animals of practical interest.

References

Constantini, F. and Lacy, E. (1981). Introduction of a rabbit β-globin gene into the mouse germline. *Nature* **294**, 92-4.

Dahlqvist, A. (1983). Digestion of lactose. In *Milk intolerances and rejection* (ed. J. Delmont) pp. 11-6. S. Karger, Basel.

Delmont, J. (ed.). (1983). Is lactose-hydrolyzed milk usable and useful? *Milk intolerances and rejection*. pp. 57-109. S. Karger, Basel.

——, Diez, J.L., Poirée, J.C., and Sudaka, P. (1983). Galactose tolerance in alcoholic liver diseases. In *Milk intolerances and rejection* (ed. J. Delmont). pp. 90-6. S. Karger, Basel.

Flatz, G. (1983). Genetics of the human adult lactase polymorphism. In *Milk intolerances and rejection* (ed. J. Delmont) pp. 27-34. S. Karger, Basel.

Fox, P.F. (ed.) (1982). *Developments in Dairy Chemistry, Vol. 1. Proteins.* Applied Science Publishers, London and New York.

Gekas, V. and Lopes-Leiva, M. (1985). Hydrolysis of lactose: a literature review. *Process. Biochem.* **20**, 2-12.

Gordon, J.W. and Ruddle, F.H. (1985). DNA-mediated genetic transformation of mouse embryos and bone-marrow — A review. *Gene* **33**, 121-36.

Hambraeus, L. (1982). Nutritional aspects of milk proteins. In *Developments in Dairy Chemistry. Vol. 1. Proteins* (ed. P.F. Fox) pp. 289-313. Applied Science Publishers, London and New York.

Hammer, R.E., Pursel, V.G., Rexroad, C.E. Jr., Wall, R.J., Bolt, D.J., Ebert K.M., Palmiter, R., and Brinster, R.L. (1985). Production of transgenic rabbits, sheep and pigs by microinjection. *Nature* **315**, 680-3.

Laporte, D.C. 1984. Anti-sense RNA: a new mechanism for the control of gene expression. *Trends in Biological Sciences* **9**, 463.

MacDonald, I. and Williams, C.A. (1983). Galactose intolerance in healthy subjects. In *Milk intolerances and rejection* (ed. J. Delmont) pp. 77-82. S. Karger, Basel.

Marshall, K.R. (1982). Industrial isolation of milk proteins: whey proteins. In *Developments in Dairy Chemistry. Vol. 1. Proteins* (ed. P.F. Fox) pp. 339-73. Applied Science Publishers, London and New York.

Melton, D.A. (1985). Injected anti-sense RNAs specifically block messenger RNA translation *in vivo*. *Proc. Nat. Acad. Sci. USA.* **82**, 144-8.

Mepham, T. B. (ed.) (1983). *Biochemistry of Lactation.* Elsevier, Amsterdam and New York.

Mercier, J. C. (1981). Phosphorylation of caseins. Present evidence for an amino acid triplet code post-translationally recognized by specific kinases. *Biochimie* **63**, 1–17.

Moneret-Vautrin, D. A. (1983). Hypersensitivity to milk in adults: present aspects. In *Milk intolerances and rejection* (ed. J. Delmont) pp. 138–41. S. Karger, Basel.

Morr, C. V. (1982). Functional properties of milk proteins and their use as food ingredients. In *Developments in Dairy Chemistry. Vol. 1.. Proteins* (ed. P. F. Fox). pp. 375–99. Applied Science Publishers, London and New York.

Muller, L. L. (1982). Manufacture of casein, caseinates and co-precipitates. In *Developments in Dairy Chemistry. Vol. 1. Proteins* (ed. P. F. Fox) pp. 315–37. Applied Science Publishers, London and New York.

Palmiter, R. D., Brinster, R. L., Hammer, R. E., Trumbauer, M. E., Rosenfeld, M. G., Birnberg, N. C., and Evans, R. M. (1982). Dramatic growth of mice that develop from eggs microinjected with metallothionein-growth hormone fusion gene. *Nature* **300**, 611–5.

Richmond, M. L., Gray, J. I., and Stine, C. M. (1981). Beta-galactosidase: review of recent research related to technological applications, nutritional concerns, and immobilization. *J. Dairy Sci.* **64**, 1759–71.

Rosenberg, U. B., Preiss, A., Seifert, E., Jäckle, H., and Knipple, D. C. (1985). Production of phenocopies by Krüppel antisense RNA injection into *Drosophila* embryos. *Nature* **313**, 703–6.

Rubenstein, J. L. R., Nicolas, J. F., and Jacob, F. (1984). L'ARN non sens (nsARN): un outil pour inactiver spécifiquement l'expression d'un gène donné *in vivo. C. R. Acad. Sci. Paris, t.299, Série III*, n°8, 271–4.

Saperstein, S. (1974). Immunological problems of milk feeding. In *Lactation. A comprehensive Treatise. Vol. 3. Nutrition and Biochemistry of Milk/Maintenance* (eds. B. L. Larson and V. R. Smith). pp. 257–80. Academic Press, New York and London.

Schmidt, D. G. (1982). Association of caseins and casein micelle structure. In *Developments in Dairy Chemistry. Vol. 1. Proteins* (ed. P. F. Fox) pp. 61–86. Applied Science Publishers, London and New York.

Skovbjerg, H., Danielsen, E. M., Noren, O., and Sjöström, H. (1984). Evidence for biosynthesis of lactase-phlorizin hydrolase as a single-chain high-molecular weight precursor. *Biochim. Biophys. Acta* **798**, 247–51.

Touchberry, R. W. (1974). Environmental and genetic factors in the development and maintenance of lactation. In *Lactation. A comprehensive treatise Vol. 3. Nutrition and Biochemistry of milk/Maintenance* (eds. B. L. Larson and V. R. Smith) pp. 349–82. Academic Press, New York and London.

Travers, A. (1984). Gene expression. Regulation by anti-sense RNA. *Nature* **311**, 410.

Weintraub, H., Izant, J. G., and Harland, R. M. (1985). Anti-sense RNA as a molecular tool for genetic analysis *Trends in Genetics* **1**, 22–5.

Discussion

Hill asked which approach, among those presented, was being pursued at Jouy-en-Josas. Mercier said his group has already cloned and almost completely sequenced the bovine α-lactalbumin gene and a joint research proposal 'Germline manipulation of livestock: production of low-lactose milk' with ABRO has been submitted to the EEC. Gibson wondered if the double-stranded RNA formed in the cell could produce anti-viral activity in the animals involved. Mercier considered this as a remote prospect since only short lengths of anti-sense RNA complementary to the 5' untranslated region of mRNA appear to be necessary for inhibition. However, this cannot be assessed without doing the experiment. Also the removal of lactose may affect the water content and yield of milk. King pointed out that the sealion produces milk with no lactose.

Hill wondered if these experiments would be done first in sheep or cattle. Mercier said the experiments would be performed first in mammalian cell lines and mice to assess the basic feasibility of both approaches. Given a positive result, subsequent steps will involve the creation of transgenic sheep and, ultimately, transgenic cows. It is likely that the same engineered gene may work in sheep, goat, and cow as the three species are phylogenetically closely related.

Gupta asked how many copies there are of the casein genes, but it is likely that there is one copy of each according to former studies of allelic segregations in the progeny of bulls heterozygous at casein loci. Evidence for the occurrence of pseudogenes is lacking. Di Berardino asked if all the casein genes are on one chromosome. The four casein genes are clustered on the same chromosome in the bovine species as deduced from the aforementioned segregation data. The four casein genes have been tentatively assigned to the q arm of metacentric chromosome 2 in the sheep as determined by *in situ* hybridization. Murine casein genes are located on chromosome 5.

14

Physiological and biochemical indicators of growth and composition

E. Müller

Abstract

The effects of selection for lipogenic enzymes in backfat of pigs on some carcass traits are analyzed. The selection criterion used was low (E^--line) or high (E^+-line) activity of NADPH-generating enzymes (sum of G-6-P-DH, 6-P-G-DH, NADP-MDH, NADP-ICDH) in backfat. A third line has been selected for low backfat thickness based on ultrasonic measurement (U^--line); there is also a control line (K-line). The selection was very successful. After eight generations of selection for high or low enzyme activity the difference between both selection lines was found to be about 2.5 phenotypic standard deviations in the selection criterion enzyme activity and about 3.6 phenotypic standard deviations in the correlated trait backfat thickness with the higher level in the E^+-line. Concerning carcass traits there are highly significant differences between the lines with greater fat content in pigs selected for high enzyme activity. In all measured traits related to adipose tissue deposition the levels of the K-line pigs are between those of the E^--line and E^+-line. In the U^--line nearly all parameters show similar levels to those of the E^--line pigs, with a tendency to lower fat content, higher meat content, and inferior meat quality.

Introduction

A number of physiological and biochemical indicators of growth have been described in the literature.

These indicators are mainly hormones and enzymes, described in a large number of different animals and tissues (literature see Etherton and Kensinger 1984). I would like to restrict my part in this seminar to the metabolism of adipose tissue in pigs. Investigations of adipose tissue in pigs have some advantages: fat metabolism is well known, repeated sampling in the fattening period by means of biopsy is feasible and the control of fat deposition in the live animal with ultrasonic measurements as well as the control of body composition in the carcass is practicable (Müller and Rogdakis 1985).

NADPH-generating enzymes in adipose tissue as indicators for fat growth and fat deposition

Function of NADPH-generating enzymes in synthesis of fat

It has been established that the production of reducing equivalents in the form of reduced nicotinamide-adenine-dinucleotide-phosphate (NADPH) is required for the synthesis of fatty acids (Langdon 1957). The pathways of NADPH production were traced quantitatively in isolated fat cells of rat adipose tissue by Kather and Brand (1975). The authors found that about 50–60 per cent of the reducing equivalents are generated during the initial steps of the pentose-phosphate pathway catalyzed by the enzymes glucose-6-phosphate-dehydrogenase (G-6-G-DH; E.C. 1.1.1.49) and 6-phosphogluconate-dehydrogenase (6-P-G-DH; E.C. 1.1.1.44). The remaining demand for NADPH is mainly provided by the enzymatic interplay within the citrate-malate-pyruvate cycle (Flatt 1970), where NADP-malate-dehydrogenase (NADP-MDH; i.e. malic enzyme E.C. 1.1.1.40) is the NADPH-generating enzyme (Fig. 14.1). NADPH-production for fatty acid synthesis is, according to Lowenstein (1961), also to be ascribed to NADP-dependent isocitrate-dehydrogenase (NADPH-ICDH, E.C. 1.1.1.42).

Many experiments, predominantly in the rat, have elucidated that a close relationship exists between the activity of these enzymes and the intensity of lipogenesis (e.g. Gibson *et al.* 1972). In the pig nearly all *de novo* fatty acid synthesis occurs in adipose tissue (O'Hea and Leveille 1969). Evidence of the relationships between the dehydrogenase activities and the rate of lipogenesis in porcine adipose tissue was given by the investigations of several authors (see Strutz and Rogdakis 1979; Müller and Rogdakis 1985).

Requirements of a biochemical indicator

For a biochemical trait to be used as a marker for a production trait it must have some special qualities. An enzyme can only be a marker if it has regulatory function in metabolism. Furthermore, the following demands have to be fulfilled: easy and exact measurement of enzyme activity and an adaptive reaction to modifications of metabolism. This means for lipogenic enzymes in pigs: high activity in obese pigs, in non-starved animals as well as in those tissues with a high potency for fatty acids synthesis (e.g. adipose tissue), respectively low enzyme activity in lean pigs, during starvation, and in muscle or liver tissue. There is a necessity for a high correlation between enzyme activity and the degree of fatness in pigs, and an appropriate proportion of the phenotypic variance must be additive genetic variance (Fewson *et al.* 1983). A great number of experiments have been performed concerning NADPH-generating enzymes (see Müller and Rogdakis 1985). Nearly all results confirm the demands mentioned above.

The topic of the following section is the description of a selection experi-

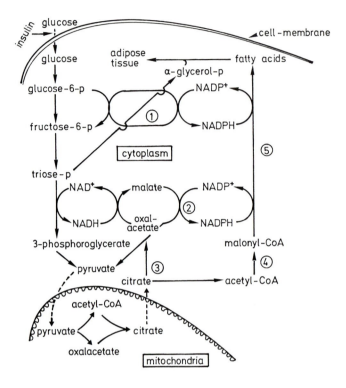

① pentosephosphatecycle
② NADP - malate - dehydrogenase
③ ATP - citrate-lyase
④ acetyl-CoA -carboxylase
⑤ fatty -acid-synthetase

Fig. 14.1.   Schematic pathway of the fatty acid synthesis

ment in pigs. The selection criterion is the activity of NADPH-generating enzymes in backfat.

Selection experiment for NADPH-generating enzymes in pigs

Materials and methods

From a basic population of German Landrace pigs the following selection lines were established.

— E⁻-line: selected for low activity of NADPH-generating enzymes in backfat (logarithm of the sum of the four enzymes G-6-P-DH,

6-P-G-DH, NADP-MDH, NADP-ICDH from two biopsies at 90 and 120 days of age).
— E⁺-line: selected for high activity of NADPH-generating enzymes.
— U⁻-line: selected for low backfat thickness at a weight of 85 kg, based on ultrasonic measurement.
— K-line: pigs are not selected.

In each generation in the E⁻-, E⁺-, and U⁻-line some 80 male and 60 female pigs are tested, and about 8 male and 16 female animals are selected. In the K-line, piglets are randomly selected (for further details on the experimental design see Rogdakis 1982).

Unfortunately, it was not possible until generation 8 in the U⁻- and K-line to measure enzyme activity. However, in the ninth generation, started a few months ago, we are measuring it in all four selection lines. Furthermore, at present female pigs and castrated boars from all four lines are being fattened up to 90 kg live weight and, after slaughtering, carcass traits are being measured.

Tissue sampling and enzyme assays are performanced as described by Strutz and Rogdakis (1979). Reference unit for enzyme activity is soluble protein (log nmol NADPH/min/mg protein).

Results and discussion

Direct selection response on enzyme activity in the E⁻- and E⁺-lines
Figure 14.2 shows the LSQ-means of enzyme activity in the E⁻- and E⁺-lines from generation 0 to 8, expressed in units of the average phenotypic standard

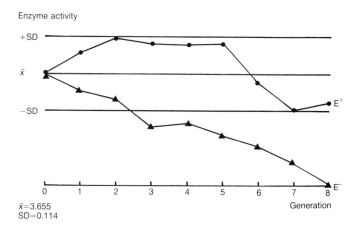

Fig. 14.2. LSQ-means of enzyme activity in E⁻- and E⁺-line from generations 0 to 8.

deviation in the population. The difference between the two lines in the 8th generation amounts to about 2.5 phenotypic standard deviations. Whereas the sum of enzyme activity in the E⁻-line in all generations show decreasing levels, selection on increasing enzyme activity (E⁺-line) has shown hardly any response since the second generation.

Correlated selection response on backfat thickness in the E⁻- and E⁺-line
Figure 14.3 shows the LSQ-means of backfat thickness in the E⁻- and E⁺-lines from generation 0 to 8, at an age of 148 days, expressed in units of the average standard deviation in the population. The difference between the two lines in the eighth generation amounts to about 3.6 phenotypic standard deviations. Both lines show an increasing differentiation in backfat thickness.

The increasing backfat thickness in the E⁺-line in all generations seems to be in contradiction to the enzyme activity in this line. The discussion about this result can only take place after completion of the 9th generation, taking into consideration the levels of enzyme activity and backfat thickness of the control line.

Comparison between selection for enzyme activity and selection for backfat thickness (preliminary results)
As mentioned above, we have for a few months now been measuring in the pigs of all four selection lines of the ninth generation enzyme activity, backfat thickness and various carcass traits. The preliminary results from about 100 pigs from all four lines and two sexes (all told some 600 pigs are to be slaughtered) are summarized in Table 14.1.

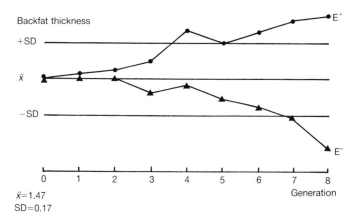

Fig. 14.3. LSQ-means of backfat thickness in E⁻- and E⁺-line from generations 0 to 8.

Table 14.1. Ranking of the selection lines, based on their enzyme activity and carcass traits

Parameter	Rank of the lines
Enzyme activity in biopsy samples	
(selection criterion in the E^--line and E^+-line)	$U^- \leq E^- < K < E^+$
Backfat thickness at an age of 148 days and 85 kg	
of weight	$U^- \leq E^- < K < E^+$
Carcass traits	
backfat thickness	$U^- \leq E^- < K < E^+$
fat cuts (%)	$U^- \leq E^- < K < E^+$
lean cuts (%)	$U^- \geq E^- > K > E^+$
ratio fat: meat	$U^- \leq E^- < K < E^+$
area of M. long. dorsi	$U^- \geq E^- = K = E^+$
Enzyme activity	
in backfat	$U^- \leq E^- < K < E^+$
in leaf	$U^- \leq E^- < K < E^+$
pH_{45} and Göfo-value	$U^- \leq E^- = K = E^+$

In all measured traits related to adipose tissue deposition the levels of the K-line pigs are between those of the E^-- and E^+-lines. As for each of M. long. dorsi, pH_{45}, and Göfo-value, these traits show no significant differences between the E^--, and E^+-, and K-line pigs. It seems that selection for low or high activity of lipogenic enzymes does not influence the area of M. long. dorsi or meat quality.

In the U^--line nearly all parameters show similar levels to that of the E^--line pigs, with a tendency to lower fat content, higher meat content and inferior meat quality. The different selection response in the four lines should be discussed when data from more animals are available.

Conclusions from the preliminary results of the selection experiment

The aforesaid results of the selection experiment with pigs and the results from other species (e.g. from rats Gibson *et al.* 1972) suggest the existence of a common genetic regulation of the whole lipogenic enzyme apparatus which is probably achieved by a regulatory event causing the simultaneous synthesis of lipogenic enzymes. These assumptions could be confirmed by the results from Sturm *et al.* (1976). They made parallel measurements of NADPH-generating enzymes and acetyl-coA-carboxylase (CBX, E.C. 6.4.1.2.) in the adipose tissue of pigs. The highly significant phenotypic correlation coefficients between CBX and the NADPH-generating enzyme levels varied between 0.77 and 0.80.

It can be assumed that in pigs the regulation of lipogenic enzymes is varied by endocrine influences, especially by the hormone insulin (e.g. Etherton and

Kensinger 1984). In our selection experiment we measured the insulin secretion of about 1000 pigs in generations 5–7 in the E⁻-, E⁺-, and U⁻-lines. The plasma insulin-levels, induced by feeding the pigs for 45 min after a fasting period of about 16 h, were very different. The animals of the U⁻- and E⁻-lines had an insulin level of about 100 μU/ml, the insulin level of the E⁺-line-pigs was about 200 μU/ml (Müller and Mailänder, in preparation). However, there was no significant difference in feed intake within these 45 min. This result could be therefore considered as an effect on selection.

For interpretation of all these results further investigations are necessary. We expect that the final results after slaughtering the animals from the ninth generation will give a little more insight into the control of fat accretion in pigs.

References

Etherton, T. D. and Kensinger, R. S. (1984). Endocrine regulation of fetal and post-natal meat animal growth. *J. Anim. Sci.* **59**, 511–27.

Fewson, D., Müller, E., and Rogdakis, E. (1983). Enzymaktivitäten als Selektionskriterium in der Schweinezucht — Modellversuche zur biochemischen Haustiergenetik. In *Forschung in der Bundesrepublik Deutschland.* pp. 415–23. Hrsg. DFG, Verlag Chemie, Weinheim.

Flatt, J. P. (1970). Energy metabolism and the control of lipogenesis in adipose tissue. In *Adipose tissue* (eds. R. Levine and E. F. Pfeiffer) pp. 93–101. Georg Thieme and Academic Press, Stuttgart/New York/London.

Gibson, D. M., Lyons, R. T., Scott, D. F., and Muto, Y. (1972). Synthesis and degradation of the lipogenic enzymes of rat liver. *Adv. Enz. Reg.* **10**, 187–204.

Kather, H. and Brand, K. (1975). Origin of hydrogen required for fatty acid synthesis in isolated rat adipocytes. *Arch. Biochem. Biophys.* **170**, 417–26.

Langdon, R. G. (1957). The biosynthesis of fatty acids in rat liver. *J. Biol. Chem.* **226**, 615–29.

Lowenstein, J. M. (1961). The pathway of hydrogen in biosynthesis. II. Extramitochondrial isocitrate dehydrogenase. *J. Biol. Chem.* **236**, 1217–9.

Müller, E., Mailänder, C., and Niebel, E. Selection for the activity of NADPH-generating enzymes in backfat of pigs. V. Population analysis of plasma-insulin secretion and some blood parameters. (Submitted to *Z. Tierzüchtung und Züchtsungsbiologie.*)

Müller, E. and Rogdakis, E. (1985). Genetische Regulation des Fettstoffwechsels beim Schwein. In *Methodische Ansätze in der Tierzüchtung.* pp. 7–28. Hrsg. S. Scholtyssek, Hohenheimer Arbeiten, **131**, Tierische Produktion, Verlag Ulmer Stuttgart.

O'Hea, E. K. and Leveille, G. A. (1969). Influence of fasting and refeeding on lipogenesis and enzymatic activity of pig adipose tissue. *J. Nutr.* **99**, 345–52.

Rogdakis, E. (1982). Selektion nach der Aktivität NADPH-liefernder Enzyme im Fettgewebe des Schweines. I. Versuchsfrage, Versuchsanlage und erste Ergebnisse. *Z. Tierz. Züchtgsbiol.* **99**, 241–52.

Strutz, Ch. and Rogdakis, E. 1979. Phenotypic and genetic parameters of NADPH-generating enzymes in porcine adipose tissue. *Z. Tierz. Züchtgsbiol.* **96**, 170–85.

Sturm, G., Rogdakis, E., Strutz, Ch., and Siebert, G. (1976). Lipogenic enzyme activity in relation to fat accumulation. *Nutr. Metabol.* **20**, 65.

Discussion

It was established that selection had been made in both sexes and in all lines, and the same selection intensity had been used. The starting phenotypic correlation between enzyme activity and fatness was of the order of 0.65. The method of adding together the activity for enzymes was questioned, but it was considered that, in the absence of more detailed information, adding them together would be one way of compensating for high activity in one and low activity in another. Future selection might in fact be with the same criterion. Halothane frequencies had been measured in the different lines and it was found that the lowest incidence was in the control line and the highest in the U-line. Some PHI differences had also been found between lines.

Questioned about the practical implications of the selection result when direct selection on low backfat had been more effective, Müller agreed with some of the comments, but said that there may not have been such big changes in meat quality and also that the enzyme method provided a different way of attacking the question of lipolysis. Smith suggested that one should not consider one method better than another, but consider combining them together in an index, which should be better than either individually.

Kanis remarked that Sejrsen had suggested that measurement of enzyme levels was unlikely to be useful and yet the present experiment had produced a useful answer. It was agreed that adipose tissue was different in that activity was being measured in the relevant tissue and not in blood. It also made sense in that there were more limited metabolic pathways in fat production than in the totality of milk production. In response to a question about the applicability to other species, it was stated that examinations were now being made with both cattle and sheep.

15

Retroviral vectors for the production of transgenic animals

Helen Sang

Note: This chapter consists of an abstract and discussion only, as it represents a paper read at short notice in the Seminar to substitute for a speaker who was unable to attend.

Abstract

Although direct injection of DNA into the nuclei of single cell embryos has been successful in the production of transgenic mice, the method has some disadvantages when its application to farm animals is considered. The major disadvantage of the direct injection technique is its low efficiency, i.e. the relatively large number of embryos that are injected for every transgenic animal that is produced. In farm animals, the difficulty and expense involved in obtaining large numbers of fertilized eggs and of performing the manipulations are much greater than in the mouse. A suitable vector system to increase the frequency of transformed embryos and to decrease the complexity of manipulations would be a great advantage. An ideal vector would need:

(1) to be easy to introduce into the experimental animal;
(2) to integrate into the genome, preferably at low copy number;
(3) to enter germ line cells; and
(4) to be able to carry foreign DNA sequences.

The most obvious candidates for development as vectors are the C-type retroviruses. These viruses already carry out the functions described above as part of their normal life cycle.

Retroviruses have an RNA genome encapsulated in a protein coat. On infection of a cell the RNA is reverse-transcribed into DNA by a virus-encoded polymerase. This results in linear DNA molecules with long terminal direct repeats (LTRs) at either end. These molecules then circularize. A viral endonuclease cuts the circles between the two repeats and the viral DNA integrates, apparently at random, into the host chromosomes. Usually, only one or a few copies of a retrovirus are found to be integrated in any particular cell. After integration the viral DNA is transcribed to give the mRNAs that code for the viral gene products, and full length copies of the viral genome. Two copies of the RNA genome then assemble with the virus core and envelope proteins to produce an infectious virus particle.

Retroviruses infect cells and integrate at low copy number into a host genome and are therefore suitable to be adapted for use as vectors. Two component systems are being developed. The vector itself is deleted for the viral coding regions but retains sequences required in cis for replication and transcription. Foreign DNA can be inserted between the LTRs. The protein components of the virus are provided in trans to package the defective vector genome. This can be done by superinfection with a healthy or helper virus, but a very low titre of vector virus particles is produced by this method. A recent improvement is the development of helper cell lines which produce all the viral proteins, but no active helper virus. Pure stocks of infective vector virus particles can therefore be obtained. Sophisticated vectors are now being developed. They usually carry an E. coli origin of replication for propagation of the vector in bacteria, and a selectable marker gene for positive selection in both bacteria and eukaryotic cells. The vectors also contain a unique restriction enzyme site, between the LTRs, for the introduction of cloned genes of interest.

The final problem involved in utilizing retroviral vectors is their introduction into the genome of an experimental animal. They have been used successfully to infect cells in tissue culture efficiently. These cells are then introduced into a host animal. This approach has not as yet produced animals with transformed germ cells, but methods are being introduced to circumvent this problem. Several other difficulties remain to be overcome. The levels of expression of cloned genes introduced in retroviral vectors tend to be lower than normal. Another possible drawback is that the amount of DNA a retrovirus can carry is probably limited. Retroviral vector systems are complex and still in the development stage, but their use may considerably increase the efficiency and decrease the difficulties involved in producing transgenic farm animals.

Discussion

Robertson asked if the retrovirus used might not leave the site of insertion as easily as it enters. Sang replied that it would probably not since mutations caused by the insertion of retroviruses are very stable. Brem wondered whether retroviruses are known in farm species, but little is known yet though it is likely that they exist in all species. The host range may be very broad at least under experimental conditions, e.g. some experiments in mice have used manipulated chicken retroviruses. Recombination between natural and introduced retroviruses is unlikely to be dangerous. Those artificially introduced will be highly modified. Lovell-Badge agreed that new retrovirus vectors even have the 5″ LTR deleted to avoid interference with transcription of inserted genes.

Gibson asked if retrovirus vectors are used with embryonic carcinoma cells, how are they then put into germ cells? Lovell-Badge said that the frequency of germ line chimerism is about 30 per cent in males in chimeras made with E.K. cells.

Hill suggested that there are particular problems in genetic manipulation of poultry, and Sang agreed, and pointed out that the large ovum makes direct injection difficult. It is, therefore, more important to improve efficiency of integration.

16

Criteria identifying genetic merit for milk production

K. Sejrsen and P. Løvendahl

Abstract

Several hormones and metabolites seem to have potential as indirect selection criteria. However, a single hormone or metabolite cannot give a reliable estimate of genetic merit for milk production, but an index of several hormones and metabolites is likely to be of value. The decision on which parameters to include in the physiological index should be based on a sound knowledge of the physiological basis for genetic differences. The parameters in the index should be measured after relevant physiological stimuli and the physiological differences between the non-lactating and the lactating state should be kept in mind when the data are interpreted. The changes in the hormone and metabolite levels that will occur when the index is used as a selection goal should be physiologically meaningful and not give rise to other negative effects. In the long run it is important to ensure a balanced improvement in the biological components limiting milk yield.

Introduction

Lactation is an integrated part of the reproductive cycle of all mammals ensuring the nurture of the offspring. Milk production is therefore an important trait in most domestic animals. Milk, however, is only the primary product in dairy cattle, sheep, and goats, and therefore the primary selection goal only in these species.

Milk production is a sex-limited trait only expressed in the mature female. Therefore, the evaluation of the breeding value of potential breeding bulls cannot be performed on the bull itself, but must be based on the performance of a group of daughters. This results in a long generation interval, since it takes 6–7 years from when the bull is born to when its breeding value for milk production is known, and a decision on whether or not to use its semen can be made. The fact that the breeding value can only be determined by a progeny test also means that a considerable part of the cows in a population must be mated to unproven bulls, some of which naturally have low breeding value.

This inherent limitation of genetic progress, due to the long generation interval and the use of unproven bulls with low breeding value, could be overcome if it was possible to evaluate breeding values of potential breeding bulls on the bulls themselves early in life. The possible extra responses from using indirect predictors alone or in combination with merit can be substantial (Walkley and Smith 1980; Christensen and Liboriussen 1985; Smith 1981). The size of the extra response depends on the relationship of the predictor with milk yield and the heritability of the predictor.

The present knowledge on indirect predictors is not sufficient for them to be used in practical breeding schemes, but several papers, that have appeared in recent years, give reason to optimism. In this paper we will attempt to give a brief review of the state of the art and to point to areas for future research.

Types of criteria investigated

Much work has been carried out in attempts to find reliable criteria identifying genetic merit for milk production. The criteria investigated include predictors with simple as well as quantitative inheritance (Kiddy 1979; Gahne 1982).

Blood groups, etc.
Several predictors with simple inheritance, i.e. blood groups, have been shown to be related to milk production. However, selection on blood groups and probably all other predictors with simple inheritance will, for theoretical reasons, only lead to a limited improvement of breeding value (Neimann-Sørensen and Robertson 1961). Predictors with simple inheritance are thus unlikely to be useful as indirect selection goals unless they represent a loci with major effect on merit. However, no major loci affecting milk production has been found so far.

Enzymes
Milk yield is a quantitative trait regulated by many genes. The yield depends on the secretory capacity of the mammary glands and the flux of nutrients through many biochemical reactions. The reactions are organized in metabolic pathways existing in many tissues. Each biochemical reaction is regulated by specific enzymes, that again are regulated by specific genes. The activity of a given enzyme is therefore a direct measure of the activity of the gene regulating the reaction and several investigators have found relationships between enzyme concentrations in blood and milk production (Kiddy 1979; Gahne 1982; Adam 1983; Graf 1984).

It is, in spite of the observed relationships with milk yield, unlikely that serum concentration of enzymes will be of any use as indirect selection criteria for milk production. There are several reasons for this. First, it is difficult to imagine that milk production is limited by one or just a few rate limiting enzymes. Secondly, will the effect of increasing the activity of a rate limiting enzyme rapidly diminish, because other enzymes become limiting? (Kascer and Burns 1979; cf. Land 1981). The third reason that serum concentrations of enzymes are unlikely to become useful, is that the enzyme concentrations in the circulation are more an indication of a metabolic stress (Graf 1984) than a reflection of the activity of the enzymes in the tissues where they act. Finally, the value of most enzymes as selection criteria is limited because they, as the blood groups, show only simple inheritance (Neimann-Sørensen and Robertson 1961). Enzymes will therefore probably only be of importance as indirect selection goals for milk production if a relationship to a major loci of importance for milk production is discovered.

Hormones

The flux of nutrients through the metabolic pathways of the various tissues of the body is regulated and co-ordinated by the endocrine system. The endocrine system also plays an essential role in the regulation of mammary development and function.

Hormones are therefore obvious candidates as predictors of merit for milk production and there is a growing amount of evidence supporting this.

Hart *et al.* (1978) and Falconer *et al.* (1980) showed that serum concentrations of the metabolic hormones growth hormone, insulin, thyroxine, prolactin, and somatomedin were different in Holstein and Holstein × Hereford cows. Several subsequent reports have shown that there also are differences in hormone concentrations of cows with different breeding value within a breed (Davey *et al.* 1983; Flux *et al.* 1984; Barnes *et al.* 1985; Bonczek *et al.* 1985; K. Sejrsen unpublished data).

It can be argued that the endocrine differences are caused by differences in energy balance (Bauman and Currie 1980; Bauman *et al.* 1985) and therefore possibly not the physiological basis of the genetic difference. However, they do reflect that animals with different genetic merit partition nutrient differently, and the observed differences in hormones do, in most cases, agree with the known effects of the hormones when administered. This is true at least for insulin and growth hormone (Kronfeld *et al.* 1963; Bauman and McCutcheon 1985).

From a breeding standpoint it is of little value that genetic merit can be identified by hormones measured in lactating cows. The aim naturally is to be able to measure the differences in the bulls.

Joakimsen *et al.* (1971), and subsequently Sørensen *et al.* (1981), have shown a relationship between thyroxine degradation rate measured in bulls and the milk yield of their daughters. Also other indicators of the thyroid system have been found to be related to breeding value for milk yield (Graf & Grosser 1979; Land *et al.* 1983; Sejrsen *et al.* 1984). Several report also indicate that differences between genetic lines can be detected in blood levels of insulin after feeding, after infusion of metabolite or during fasting (Land *et al.* 1983; Gränzer *et al.* 1983; Schwab *et al.* 1984; Sejrsen *et al.* 1984). Barnes *et al.* (1985) found increased serum growth hormone after feeding in heifers from cows selected for high merit, whereas Land *et al.* (1983) and Schwab *et al.* (1984) found lower growth hormone levels in bull calves and bulls with high merit during fasting. Barnes *et al.* (1985) also found differences in serum prolactin after feeding and insulin administration.

Metabolites

The corner stone in the endocrine regulation of metabolism is a biological need (Roth and Grunfeld 1981), that often manifests itself as a change in circulating levels of one or more metabolites. The changes in the metabolite concentrations result in a change in the synthesis and secretion of one or more hormones. The biological effects of the released hormones result in a fulfilment of the biological need and the metabolite as well as the hormone levels return to normal. An obvious example is the insulin response to a glucose load. It illustrates the inter-relationship between hormone and metabolite levels and explain why metabolites as well as hormones have potential as criteria identifying genetic merit for milk production.

The potential of metabolites as indirect predictors of merit for milk yield is supported from reports by Tilakaratne *et al.* (1980), Almlid *et al.* (1982), Sejrsen *et al.* (1984), and Barnes *et al.* (1985). Especially blood concentrations of urea, free fatty acids, and possibly ketones seem to have potential as predictors.

Discussion on future research

The existing body of evidence indicates that a reliable criteria identifying genetic merit for milk production can be developed based on measurements of hormones and metabolites. There are, however, still many questions to resolve before a criterion is developed and ready for use.

What hormones and metabolites should be used?

The biological components determining milk yield are regulated by complexes of hormones, some with opposing actions (Tucker 1981a, b; Akers 1984; Bines and Hart 1982). It is therefore unrealistic to believe that one hormone or metabolite alone can explain a major part of the genetic variation in milk yield capacity, and a reliable criteria identifying genetic merit will most likely have to be a physiological index of several hormones and metabolites. Sejrsen *et al.* (1984) and Schwab *et al.* (1984) have both observed a considerable increase in the predictive value of physiological indicators when two or more physiological measurements were used in combination.

A logical search, for hormones and metabolites to use in a physiological

index of genetic merit for milk production, has to be based on a sound knowledge of the physiological basis for genetic difference and the changes in
hormone and metabolite levels, resulting from the use of the index as
selection goal, should be physiologically meaningful and not give rise to other
negative effects. The main reason that thyroxine degradation rate is not used
as indirect selection goal, even if it has been calculated that its use could lead
to a 50 per cent increase in the annual genetic progress (Sørensen *et al.* 1981),
is that the observed relationship is in conflict with our present knowledge on
the effect of thyroid hormones. It is likely that an increase in the degradation
rate would result in a decrease in feed efficiency due to increased heat
production.

In order to avoid such unwanted effects, it could be argued that a physiological index for milk yield should include all hormones affecting the trait,
even if the accuracy of prediction is not improved by adding all of them.

How and when to measure the parameters?

Several investigations on the use of hormones as criteria identifying genetic
merit for milk production have yielded disappointing results (i.e. Tucker *et
al.* 1974; Osmond *et al.* 1981). In these investigations the endocrine status of
the bulls were based on single samples, and little or no consideration was
given to the influences of environmental factors. It is now well documented
that feeding level, time after feeding, season, temperature, and daylength, as
well as age and physiological state of the animal affect the circulating level of
hormones (Trenkle 1978; Tucker 1982). It has therefore over the last few
years become evident that animals should be compared at the same age and
physiological state, and that physiological stimuli or challenges should be
used to eliminate or minimize environmental influences (Tilakaratne *et al.*
1980; Osmond *et al.* 1981; Hart *et al.* 1981; Land *et al.* 1983; Sejrsen *et al.*
1984).

Another advantage of using physiological stimuli is that they can be used to
simulate the metabolic conditions of the cow during lactation. However, it is
important to consider possible differences between the physiological states
when the data are interpreted. This is illustrated by the findings on the effect
of breeding value on plasma urea. Tilakaratne *et al.* (1980) and Sejrsen *et al.*
(1984) found that plasma urea during fasting was highest in bull calves with
low genetic merit, which is in contrast to what have been found in cows by
Hart *et al.* (1978) and Sejrsen (1985).

Physiological stimuli can also be used to simulate specific metabolic
processes of importance for milk yield, i.e. adrenaline infusion stimulates
lipolysis, glucose infusion stimulates insulin release, and thyrotropin-
releasing factor stimulates growth hormone release. Infusion of hormones
can also give an indication of differences in tissue responsiveness to the
hormone (Bauman *et al.* 1985).

In relation to the use of physiological stimuli it is important to consider how the responses are best evaluated. Should we consider peak value, maximum changes, area under the curve, or rate of decay? Furthermore, how many samples are needed? Can they be pooled? The decisions, of course, depend on the physiology of the reaction, but factors such as repeatability, and cost of labour and chemical analyses should also be taken into account. A test such as thyroxine degradation is fairly complicated to perform and it involves the use of radioactive isotopes, which is not always acceptable. A more simple test of the thyroid system is therefore needed.

In order to be of use in breeding the parameters have to be measured before the bulls are 1 year of age. Most data indicate that the challenges should be performed as early as possible. This, however, may not be true for all parameters. Sex hormones, for example, cannot be measured until after onset of puberty. The best age to perform the tests may therefore vary from parameter to parameter. The extra information obtained from measuring at the best time for a given parameter must therefore be weighed against the extra effort.

Biological components limiting milk yield

Genes controlling milk production are common to both sexes, but several of the genes controlling mammary secretory capacity are not expressed in the male. For example, several of the hormones regulating mammary development limited to the female sex and the male sex hormone inhibit mammary development. It is therefore difficult to suggest physiological criteria in the male that can give an indication of the mammary secretory capacity of their offspring.

Indirect measures of mammary secretory capacity already exist in the female. Sweet *et al.* (1955) and Nielsen (1960) have found relationships between palpation score at 3–6 months of age and subsequent milk yield. If computer assisted tomograph scanners were large enough to hold calves, an even more reliable measure of mammary secretory capacity could be obtained (Sørensen *et al.* 1985). The genetic importance of this is limited under the present breeding schemes, but continued progress in embryo splitting and transfer techniques may make an effort in this area worthwhile in the future.

Mammary development before pregnancy is regulated by the female sex hormone oestrogen and growth hormone. We have shown that mammary development is positively related to serum levels of growth hormone (Sejrsen *et al.* 1983) and this reflects a cause and effect relationship (Sejrsen *et al.* 1986). If this also reflects a relationship between blood levels of growth hormone in bulls and the mammary development in their daughters the possibility exist that at least some indication of the daughters mammary secretory capacity can be obtained in the bull.

The work on finding indirect predictors of merit in bull calves has so far been concerned mainly with the nutrient supply, the other major biological component of milk yield. However, an unbalanced progress in the components of the traits may limit improvement (Smith 1981). Progress in mammary nutrient supply is of no use if milk yield is limited by the secretory capacity of mammary glands. It is therefore important that progress in secretory capacity is maintained at same level as progress in nutrient supply.

Genetic parameters and incorporation into the breeding plan

The research aimed at finding the right physiological criteria identifying genetic merit is probably best carried out on animals from selected lines. This means that the genetic parameters still remain to be established, when it is decided which parameters that should be included in the index. When the genetic parameters are established it has to be investigated how the index is best incorporated into the total breeding scheme and possible biological consequences on other traits have to be evaluated.

Conclusions

1. Criteria with simple inheritance are of limited use as indirect selection goal unless a loci with major effect on milk yield is identified.

2. Hormones and metabolites seem to have potential as predictors of genetic merit for milk production.

3. One hormone or metabolite cannot give a reliable estimate of genetic merit. An index consisting of several hormones and metabolites should be formed.

4. The decision on which parameter to include in the index should be based on a sound knowledge of the physiological basis for genetic differences in milk production capacity.

5. The changes in hormone and metabolite levels due to the use of the index should be physiologically meaningful and not give rise to negative effects.

6. The hormones and metabolites should be measured after relevant physiological stimuli. Single, acute samples are not adequate.

7. The physiological difference between the non-lactating and the lactating state should be kept in mind when the data are interpreted.

8. It is important, at least in the long run, to ensure a balanced improvement in the biological components limiting milk yield and to consider the biological consequences for other important traits.

9. When the physiological index is developed, the genetic parameter still remain to be established and it has to be investigated how the index is best incorporated into the total breeding scheme.

References

Adam, F. (1983). Enzymaktivitäten und Substratkonzentrationen bei Milchkühen, untersucht im Blut von monozygoten Zwillingen. *Kieler Milchn. Forsch. Ber.* **35**, 3-87.

Akers, R.M. (1985). Lactogenic hormones: Binding sites, mammary growth, secretory cell differentiation and milk biosynthesis in ruminants, *J. Dairy Sci.* **68**, 501-19.

Almlid, T., Halse, K., and Tveit, B. (1982). Effects of fasting on blood levels of acetoacetate, free fatty acids, glucose, insulin and thyroxine in bulls: differences between individuals. *Acta Agri. Scand.* **32**, 299-303.

Barnes, M.A., Kazmer, G.W., Akers, R.M., and Pearson, R.E. (1985). Influence of selection for milk yield on endogenous hormones and metabolites in Holstein heifers and cows. *J. Anim. Sci.* **60**, 271-84.

Bauman, D.E. and Currie, W.B. (1980). Partitioning of nutrients during pregnancy and lactation: A review of mechanisms involving homeostasis and homeothesis. *J. Dairy Sci.* **63**, 1514-29.

—— and McCutheon, S.N. (1985). The effects of growth hormone and prolactin on metabolism. In *Proceedings of VI International Symposium on Ruminant Physiology*, (ed. L.P. Milligam, W.L. Grovum, and A. Dobson). Reston Publ. Co., USA.

——, ——, Steinhour, W.D., Eppard, P.J., and Sechen, S.J. (1985). Sources of variation and prospects for improvement of productive efficiency in the dairy cow: A review. *J. Anim. Sci.* **60**, 583-92.

Bines, J.A. and Hart, I.C. (1982). Metabolic limits to milk production: especially roles of growth hormone and insulin. *J. Dairy Sci.* **65**, 1375-89.

Bonczek, R.R., Young, C.W., Wheaton, J.E., and Miller, K.P. (1985). Effect of selection for milk yield on plasma growth hormone concentrations of Holstein cows at two stages of lactation. *J. Dairy Sci.* **68**, 216.

Christensen, L.G. and Liboriussen, T. (1985). Embryo transfer in genetic improvement of dairy cattle. In *Exploiting New Technologies in Animal Breeding* (eds. Smith, King and MacKay), Oxford University Press.

Davey, A.W.F., Gringer, C., MacKenzie, D.D.S., Flux, D.S., Wilson, G.F., Brookes, I.M., and Holmes, G.W. (1983). Nutritional and physiological studies on differences between Friesian cows of high and low genetic merit. *Proc. N.Z. Soc. Anim. Prod.* **43**, 67-70.

Falconer, J., Forbes, J.M., Bines, J.A., Roy, J.H.B., and Hart, I.C. (1980). Somatomedin-like activity in cattle: the effect of breed, lactation and time of day. *J. Endocr.* **86**, 183-8.

Flux, D.S., MacKenzie, D.D.S., and Wilson, G.F. 1984. Plasma metabolite and hormone concentrations in Friesians cows of differing genetic merit measured at two feeding levels. *Anim. Prod.* **38**, 377-84.

Gahne, B. 1982. Use of additional traits in dairy cattle breeding. *2nd World Congr. Gen. Appl. Livest. Prod.* **V**, 387-98.

Graf, F. (1984). Of what value are blood tests (physiological characteristics) to estimate the production capacity and reliability of cattle. *Anim. Res. Dev.* **19**, 118-25.

—— and Grosser, I. (1979). Thyroxinwerte von Jung Bullen in Bezug zur Eigen-leistung und zur Tochterleistung. *Zbl. Vet.-Med. A,* **26**, 682–6.

Gränzer, W., Hahn, R., and Pirchner, F. (1983). Die Insulin-konzentration Zuchtwert. *Zuchtungskunde* **55**, 91–9.

Hart, I. C., Bines, J. A., Morant, S. V., and Ridley, J. L. (1978). Endocrine control of energy metabolism in the cow: Comparison of the levels of hormones (prolactin, growth hormone, insulin, and thyroxine) and metabolites in the plasma of high and low yielding cattle at various stages of lactation. *J. Endocr.* **77**, 333–45.

——, Morant, S. V., and Roy, J. H. B. (1981). A note on the variability of hormone concentrations in twice-weekly blood samples taken from heifers during the first 110 days of life. *Anim. Prod.* **32**, 215–7.

Joakimsen, O., Steenberg, K., Lien, H., and Theodorsen, L. (1971). Genetic relation-ship between thyroxine degradation and fat-corrected milk yield in cattle. *Acta Agric. Scand.* **21**, 121–4.

Kacser, H. and Burns, J. A. (1979). Molecular democracy: who shares the controls? *Biochemical Reviews* **7**, 1149–60.

Kiddy, C. A. (1979). A review of research on genetic variation in physiological charac-teristics related to performance in dairy cattle. *J. Dairy Sci.* **62**, 818–24.

Kronfeld, D. S., Mayer, G. P., Robertson, J. M., and Raggi, F. (1963). Depression of milk secretion during insulin administration. *J. Dairy Sci.* **46**, 559–63.

Land, R. B. (1981). Physiological criteria and genetic selection. *Livest. Prod. Sci.* **8**, 203–13.

——, Carr, W. R., Hart, I. C., Osmond, T. J., Thompson, R., and Tilakaratne, N. (1983). Physiological attributes as possible selection criteria for milk production. 3. Plasma hormone concentrations and metabolite and hormonal responses to changes in energy equilibrium. *Anim. Prod.* **37**, 165–78.

Neimann-Sørensen, A. and Robertson, A. (1961). The association between blood groups and several production characteristics in three Danish cattle breeds. *Acta Agri. Scand.* **11**, 163–96.

Nielsen, J. (1960). Måling af yverkirtlerne hos kviekalve som metode til vurdering af disses ydelsesanlæg. *Report Nat. Inst. Anim. Sci.* **321**, 1–20.

Osmond, T. J., Carr, W. R., Hinks, C. J. M., and Land, R. B. (1981). Physiological attributes as possible selection criteria for milk production. 2. Plasma insulin, tri-iodothyronine and thyroxine in bulls. *Anim. Prod.* **32**, 159–63.

Roth, J. and Grundfeld, C. (1981). Endocrine Systems: Mechanisms of disease, target cells and receptors. In: *Textbook of Endocrinology* (ed. R. H. Williams), pp. 15–72. Saunders, Philadelphia.

Schwab, M., Pirchner, F., Peeters, G., Gränzer, W., Groth, W., and Hahn, R. (1984). Physiological reactions before and after fasting in relation to dairy merit of bulls. *35th Ann. Mtg. EAAP.*

——, Huber, J. T., and Tucker, H. A. (1983). Influence of the amount fed on hormone concentrations and their relationship to mammary growth in heifers. *J. Dairy Sci.* **66**, 845–55.

——, Larsen, F., and Andersen B. B. (1984). Use of plasma hormone and metabolite levels to predict breeding value of young bulls for butterfat production. *Anim. Prod.* **39**, 335–44.

Sejrsen, K., Foldager, J., Sørensen, M. T., Akers, R. M., and Bauman, D. E. (1986).

Effect of exogenous bovine somatotropin on pubertal mammary development in heifers. *J. Dairy Sci.* **69**, 1528-1544.

Smith, C. (1981). Genetical and statistical aspects of physiological predictors of merit. Scientific Workshop, Edinburgh, Sep. 1981.

Sørensen, M. T., Kruse, V., and Andersen, B. B. (1981). Thyroxine degradation rate in young bulls of Danish dual-purpose breeds. Genetic relationship to weight gain, feed conversion and breeding value for butterfat production. *Livest. Prod. Sci.* **8**, 399-406.

——, Sejrsen, K., and Foldager, J. (1985). Prediction of parenchyma content in heifer mammary glands by computed tomography (CT). *J. Dairy Sci.* (submitted).

Sweet, W. E., Book, I. H., Matthews, C. A., and Fohrman, M. H. 1955. Evaluation of mammary gland development in Holstein and Jersey calves as a measure of potential producing capacity. *U.S. Dept Agric. Tech. Bull.* **1111**.

Tilakaratne, N., Alliston, J. C., Carr, W. R., Land, R. B., and Osmond, T. J. (1980). Physiological attributes as possible selection criteria for milk production. 1. Study of metabolites in Friesian calves of high or low genetic merit. *Anim. Prod.* **30**, 327-40.

Trenkle, A. (1978). Relation of hormonal variations to nutritional studies and metabolism of ruminants. *J. Dairy Sci.* **61**, 281-93.

Tucker, H. A. (1981a). Physiological control of mammary growth lactogenesis and lactation. *J. Dairy Sci.* **64**, 1403-21.

—— (1982b). Seasonality in cattle. *Theriogenology* **17**, 53-9.

——, Koprowski, J. A., Britt, J. H., and Oxender, W. D. (1974). Serum prolactin and growth hormone in Holstein bulls. *J. Dairy Sci.* **57**, 1092-4.

Walkley, J. R. W. and Smith, C. (1980). The use of physiological traits in genetic selection for litter size in sheep. *J. Reprod. Fert.* **59**, 83-8.

Discussion

The question was raised as to whether hormone levels after stimulus or a combination of hormone and receptor levels was more appropriate for predictive purposes. Sejrsen suggested that attention should be focussed on those measures known to be physiologically relevant and that although receptors would be included, they would not be over-emphasized as they had been at one time. The stability of predictions over a period of time was queried and it was agreed this needs to be monitored.

Hill suggested that physiologists try to have it both ways in suggesting that animals with high levels of hormone showed higher hormone activity, while those with low levels indicated high receptor activity. It was thought that increased hormone levels were the more important. Land suggested a case for more interaction between genetics and physiology so that use of both disciplines could lead to a better understanding.

Smith wondered why the curve shown, relating accuracy of prediction to the number of predictors, was not exponential, but this would only have been expected had the predictors been added in order of utility and this had in fact not been done. Cameron suggested that the first step should be to construct a genetic index and not a physiological index. Discussion led to agreement that it was more appropriate to identify relevant variables and then to derive the genetic parameters.

The question was posed whether, when studying hormone differences as predictors, we should be doing so in populations with or without major genes. It was thought that single genes were likely to be atypical and should therefore be avoided for the purposes of generalization. It was important also to avoid confusion with breed differences.

17

The Chinese prolific breeds of pigs: examples of extreme genetic stocks

P. Sellier and C. Legault

Abstract

A limited number of Chinese native breeds, essentially those of the Taihu group, exhibit exceptional prolificacy. This paper includes a survey of the evaluation work carried out in France on two Taihu breeds, i.e. Meishan and Jiaxing. Information collected so far allows us to conclude that the advantage of half-Chinese sows, as compared to European crossbred sows, is of the order of five to eight additional pigs weaned per sow per year, representing an improvement of about 30 per cent. However, the use of Chinese breeds causes great problems as regards economy and composition of gain in the growing animal: for instance, the disadvantage of 1/4 Chinese pigs amounts to around 17–19 per cent in lean tissue food conversion.

Exploiting Chinese germplasm in current European breeding programmes may be done through systematic terminal crossing involving 1/2 or 1/4 Chinese dams, or alternatively by introducing Chinese genes to new composite dam lines at a rate of 25–50 per cent. Advantages and disadvantages of each system are discussed from both genetic and organizational viewpoints. Searching for a possible major gene responsible for the high prolificacy of Taihu breeds is of interest for the future.

Introduction

Prolificacy is a trait of great importance in pig production. However, little progress in number of piglets born per litter has actually been achieved in European breeds in the last 30 years. In fact, prospects of success regarding selection for prolificacy are relatively low. It is indeed well known that only a small portion of the within-breed variation in this trait is controlled by additive gene effects, the remainder of the variation being of environmental or non-additive genetic origin (e.g. Bolet and Legault 1982; Hill and Webb 1982).

In the People's Republic of China, there exists a limited number of native breeds of exceptional prolificacy. These breeds essentially belong to the Taihu group (Meishan, Jiaxing, Fengjing and Erhualian; see Cheng 1983). It may be of value to incorporate them in breeding programmes run in

European countries, in order to make significant advances in average prolificacy of the sow population.

The purpose of this paper is (1) to give an up-to-date survey of the evaluation work carried out in France on two of the Taihu breeds, namely Meishan and Jiaxing, and (2) to briefly discuss possible routes for taking advantage of these highly prolific breeds in Europe.

Evaluation of Meishan and Jiaxing breeds in France

Reproductive traits

As far as meat-producing mammals are concerned, reproductive performance is usually assessed by numerical productivity, defined as the number of youngs weaned per breeding female and per year of presence in the herd. Sow productivity depends on several components among which litter size plays a major role.

Table 17.1.   Litter size and weight from purebred Chinese, 1/2 Chinese, 1/4 Chinese, and purebred European sows (Legault *et al.* 1984)

Genetic type of the dam	No. of litters (no. of sows)	Litter size			Litter weight (kg)	
		Total born	Born alive	Weaned (at 28 days)	At birth	At 21 days
MS	115 (35)	14.9a	14.0a	13.1a	16.2b	57.3b
JX	86 (29)	11.6b	10.8b	10.0b	9.5c	38.5c
MS × (LW or FL)	107 (42)	15.3a	14.5a	12.8a	19.3a	67.8a
JX × (LW or FL)	68 (29)	15.2a	14.7a	13.2a	15.8b	64.5a
LW × (1/2MS or 1/2JX)	63 (24)	11.5b	10.8b	9.9b	15.6b	57.6b
LW or FL	42 (22)	10.7b	10.2b	9.2b	14.7b	56.8b

MS = Meishan, JX = Jiaxing, LW = Large White, FL = French Landrace.
Means with different letters are significantly different ($P < 0.05$).

The results obtained for litter size at birth and at weaning in the French experiment (from 1980 to 1983) are summarized in Table 17.1. Pure-bred Meishan (MS) and Jiaxing (JX) sows, as well as sows from various crosses with 1/2 or 1/4 of Chinese breeding were evaluated with purebred Large White (LW) and French Landrace (FL) sows used as controls. As compared to LW or FL sows, MS sows farrowed around four additional piglets per litter, whereas prolificacy of JX sows was surprisingly the same as that of European sows. However, MS and JX breeds gave similar results in litter size when they were evaluated on the basis of F_1 sows. Emphasis is to be laid on the fact that F_1 Chinese × European sows exhibited about the same prolificacy as pure MS sows, showing that strong heterosis effects on litter size are expressed in crosses between Chinese and European breeds. Estimates of maternal heterosis reported by Legault (1985) amount to 26 and 37 per cent for litter size at birth, and 18 and 37 per cent for litter size at weaning, in LW × MS and LW × JX crosses respectively; these values are markedly higher than values of 6–8 per cent generally found in crosses between European breeds. In addition, it can be mentioned that the superiority of MS over LW sows was of the same magnitude as that reported above in a more recent comparison dealing with a new sample of Large Whites (Table 17.2); the mating design was a 2 × 2 diallel cross, with sows producing either pure- or cross-bred litters.

Table 17.2. Litter size and weight from purebred Meishan and Large White sows (J.C. Caritez and C. Legault, unpublished data)

Breed of dam	No. of litters	Litter size			Litter weight (kg)	
		Total born	Born alive	Weaned	At birth	At 21 days
MS	48	15.8a	14.4a	12.9a	15.8a	55.0a
LW	38	10.7b	10.0b	9.0b	12.9b	50.8a

Means with different letters are significantly different.

A point of great interest is to know whether the high prolificacy of those Chinese breeds is due to higher ovulation rate, or to higher embryonic survival rate, or to both of them. From data collected so far in France (Bolet *et al.* 1986), it may be suggested that the advantage of MS sows, as compared to LW sows, essentially comes from lowered embryonic mortality rate, ovulation rate being the same in the two breeds (Table 17.3). It should also be pointed out that the pattern of superiority of the so-called 'hyperprolific' sows (Legault and Gruand 1976) is quite different: the advantage of 'hyperprolific' over control LW sows is due to a steady increase in ovulation rate.

Table 17.3. Comparison of Meishan sows and control and 'hyperprolific' Large White sows in ovulation rate (OR) and embryonic mortality rate (EMR) (Bolet *et al.* 1986)

Genetic type	No. of sows		Mean ± SE	
	OR	EMR	OR	EMR (%)
MS	17	16	17.2 ± 1.2	16 ± 9
control LW	15	13	17.6 ± 0.8	26 ± 7
'hyperprolific' LW	18	17	22.9 ± 1.0	41 ± 7

The duration of the oestrus cycle in Chinese and half-Chinese gilts is not significantly different to that of European gilts, whereas the duration of heat tends to be longer than that usually found in European breeds (about 53 h), especially in MS pure breds and MS crosses (about 70–75 h), as reported by Rombauts *et al.* (1982). Post-weaning oestrus occurs about one day earlier in Chinese than in European contemporary sows. Moreover, Chinese sows, especially JX ones, have a propensity to exhibit fertile oestrus during lactation.

Table 17.4.   Age at puberty and teat number in Chinese and half-Chinese gilts (Legault and Caritez 1983)

Genetic type	Age at puberty (days)*		Teat number†	
	No. of gilts	Mean ± SE	No. of gilts	Mean ± SE
MS	36	81 ± 2	136	16.3 ± 0.1
JX	22	91 ± 2	81	19.9 ± 0.2
MS × (LW or FL)	38	87 ± 2	87	14.7 ± 0.2
JX × (LW or FL)	40	93 ± 2	80	16.7 ± 0.2

* Average age at puberty in LW and FL gilts is around 195–210 days.
†Average teat number in LW and FL gilts is near 14.

It was reported that in China puberty is attained between 2 and 4 months of age in gilts as well as in boars of the Taihu breeds (e.g. Zhang *et al.* 1983; Cheng 1983). This early sexual maturity was confirmed in the French environment (Legault and Caritez 1983; Bolet *et al.* 1985). In addition, half-Chinese gilts reached puberty at about the same age as pure-bred Chinese gilts (Table 17.4), indicating that age at first oestrus is affected to a large extent (about 30–40 per cent) by heterosis in Chinese × European crosses, as pointed out by Legault and Caritez (1983). The earlier sexual maturity of half-Chinese gilts could reasonably lead to reducing age at first farrowing by at least 1 month, as compared to current practice in Europe. Another point of interest is the higher teat number of Chinese females, especially the Jiaxing ones according to the French results (Table 17.4). This trait is transmitted in an additive way in Chinese × European crosses, as indicated by more recent data (J. C. Caritez and C. Legault, unpublished results): average teat number was 16.9, 15.3, 14.8, 19.8, 16.6, 15.4, and 13.8 in MS, 1/2 MS, 1/4 MS, JX, 1/2 JX, 1/4 JX, and LW gilts, respectively.

Production traits
Growth and carcass traits are also to be taken into account in evaluating Chinese breeds. In this field, several comparisons between crosses having varying percentages of Chinese and European genes were carried out from

1980 to 1983 in France, as reported by Legault *et al.* (1985). Pure-bred Chinese and half-Chinese slaughter pigs do not meet the current market requirements in European countries, due to excessive fatness and very poor muscle development. This precludes the use of pure Chinese sows as dams in commercial herds.

Table 17.5 gives results of a comparison between 1/4 MS, 1/4 JX and control European pigs, produced in a 2 × 3 diallel cross comprising two European breeds of sire, i.e. Large White and Belgian Landrace, and three genetic types of F_1 dams, i.e. MS × (LW or FL), JX × (LW or FL), and FL × LW.

Table 17.5. Growth and carcass traits in 1/4 Chinese and control European slaughter pigs Legault *et al.* 1985)

Trait	Genetic type		
	1/4 MS	1/4JX	European control
No. of animals (no. of dams)	317 (31)	306 (20)	85 (22)
Average daily gain (g)	790b	754c	818a
Food conversion ratio (kg feed/kg gain)	3.63b	3.74b	3.40a
Killing out percentage	77.9b	78.4a	78.3ab
Carcass length (cm)	95.8b	97.9a	97.0a
Backfat thickness rump	31.1c	29.9b	27.8a
(mm) back	27.3b	25.6a	25.1a
Estimated lean percentage (E.E.C. reference)	45.6b	45.1b	49.1a
Meat quality index (points)	86.3a	86.1a	85.8a

Means with different letters are significantly different.

The 1/4 MS and 1/4 JX pigs showed acceptable average daily gain, lower than but comparable to that of European controls. A marked heterosis effect on growth rate is likely to account for the relatively satisfactory situation of 1/4 Chinese pigs in this respect. However, these 1/4 Chinese pigs presented higher food conversion ratio and lower estimated lean percentage than controls. From the results of various comparisons between crosses in the French experiment, Legault *et al.* (1985) derived the presumed position of MS and JX breeds, as compared to the main breeds exploited in France, in killing out percentage and carcass lean percentage (fig. 17.1). The two Chinese breeds appear to be hardly inferior to LW and FL breeds as regards killing out percentage, but they would be inferior by 16–18 percentage points in carcass lean content. The latter difference is probably slightly under-estimated: the prediction equation used for estimating lean percentage was

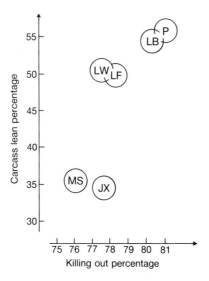

Fig. 17.1. Presumed position of Meishan (MS) and Jiaxing (JX) breeds, as compared to Large White (LW), French Landrace (LF), Belgian Landrace (LB), and Pietrain (P) breeds, in killing out percentage and estimated carcass lean percentage (E.E.C. reference).

indeed established from data collected on European pigs and perhaps gave results biased upwards when applied to pigs of Chinese breeding.

As regards meat processing ability, the present state of information suggests that MS and JX breeds do not differ to a large extent from LW breed. The meat quality index presented in Table 17.5 is intended to predict the technological yield of Paris cooked ham processing, and no difference was found between 1/4 Chinese and European pigs in this respect. However, there is evidence that half-Chinese pigs present an advantage in tenderness, juiciness, and to a lesser extent flavour of the meat, as judged by taste panels (Touraille *et al*. 1983, 1985). Higher intramuscular fat content (marbling) is probably at the origin of this advantage.

Conclusions

The above results should be interpreted cautiously due to the very limited size of the foundation samples of Chinese breeds imported to France. However, as they are in general agreement with data collected in China (Wu and Zhang 1982; Zhang *et al*. 1983; Cheng 1983, 1984), they provide a satisfactory basis for defining the possible ways of exploiting highly prolific Chinese breeds.

Information collected so far in the French experiment dealing with Taihu breeds can be summarized as follows.

(1) Genes from Meishan and Jiaxing breeds undoubtedly offer great promise regarding the improvement of litter size. Thus, the advantage of 1/2 Meishan or 1/2 Jiaxing sows, as compared to the best European genetic types of sows, amounts to five to eight additional piglets weaned per sow per year, representing an improvement of about 30 per cent in numerical productivity.

(2) At the same time, these Chinese breeds introduce great problems with respect to economy and composition of gain in the growing animal. For instance, it can be deduced from Table 17.5 that the disadvantage of 1/4 Meishan or 1/4 Jiaxing pigs, as compared to European controls, is approximately of 12–14 per cent in lean tissue growth rate and 17–19 per cent in lean tissue food conversion.

Possible routes for exploiting prolific Chinese breeds

Several approaches can be proposed for taking advantage of highly prolific breeds of the Taihu type.

Use of half-Chinese dams in a terminal crossing system

This method consists of producing F_1 dams by crossing a Chinese breed (e.g. Meishan) and a 'dual-purpose' European breed (e.g. Large White), and then to use a specialized sire breed (e.g. Pietrain) for the terminal cross. This three-way crossing would take full benefit of the maternal heterosis effect on numerical productivity and of the complementarity between sires and dams of the terminal cross. If the absence of any difference in the prolificacy of sows from the two reciprocal F_1 crosses is confirmed, it would be preferable to use the Chinese breed as the dam line of the first cross in order to lower the production cost of the F_1 gilt.

However, to use a Chinese breed in such a systematic cross-breeding scheme one must create and maintain a pure-bred Chinese nucleus of sufficient effective size, and this may presently cause problems. In addition, as regards selection in favour of production traits in this nucleus, difficulties will be encountered for accurately testing young Chinese boars on their own performance, due to very early puberty and great sexual activity.

Economic calculations should be made for judging the respective merit of this method and of the current crossing systems based only on European dam breeds. In few words, the question is to know whether the decrease in production cost of the weaned piglet, as brought by the higher numerical productivity of half-Chinese sows, is offset or not by the reduction of the gross margin from fattening in their 1/4 Chinese offspring. The answer to this question depends on the economic parameters of pig production in each country and no general answer can be given at the present time.

Use of 1/4 Chinese dams in a terminal crossing system

This method derived from the previous one consists of making a three-way cross on the maternal side of the crossing system in order to produce 1/4 Chinese sows, and then to mate these sows to specialized terminal boars. For instance, starting from Chinese females, two European breeds of the 'dual-purpose' type (e.g. Large White and Landrace) can be successively used for producing commercial gilts. This theoretically leads to exploit only one-half of the large maternal heterosis effect on litter size found in Chinese × European crosses. It may be argued that the rather disappointing prolificacy levels obtained so far in France with 1/4 Meishan or Jiaxing sows reflects this loss in heterosis. Maintaining a Chinese nucleus is needed in this system with the same problems to cope with as above.

Creation of a composite line including Chinese genes

Another solution could be the introduction of Meishan or Jiaxing genes in a European population of good merit for prolificacy in order to create a new composite dam line with a certain rate of Chinese genes. If one Chinese and one European breed equally contribute to the genetic make-up of the composite line, only one half of the F_1 maternal heterosis effect on litter size, as predicted from the simple dominance model of heterosis, will be retained in the F_2 and subsequent generations. 'Epistatic recombination losses' could also negatively affect reproductive performance in the composite line.

This method is relatively easy to set up on practical grounds: for instance, semen from Chinese breeds can be imported once for making the F_1 foundation cross of the composite line. As regards growth and carcass traits, an increase in genetic variance and therefore in selection response may be expected in such a composite line, whereas boar performance-testing will hopefully be easier to conduct than in a pure-bred Chinese nucleus. However, the extent to which intense selection on growth and carcass traits will affect reproductive performance of the new line remains unknown.

Searching for a 'major gene'

The origin of the high prolificacy of Taihu breeds is a question of particular interest from a purely genetic viewpoint. Long-time selection on litter size could have led to accumulate a large number of favourable genes each with small or moderate effects. As higher embryonic survival seems to be the main explanation of high prolificacy in the Taihu breeds, the phase of close inbreeding which has probably taken place during the development of these local breeds could also have contributed to eliminate many deleterious genes responsible for embryonic mortality.

However, that a major gene is involved is another attractive hypothesis. Several methods have been proposed for detecting individual genes with large effects on quantitative traits (see Roberts and Smith 1982 for a review).

Experimental work required for testing this hypothesis will be considerable even though the effect of the postulated major gene affecting prolificacy, if it does exist, is likely to be of appreciable magnitude in terms of standard deviation units of the trait. Such a gene would be, of course, an excellent candidate for study in the field of genetic engineering.

References

Bolet, G. and Legault, C. (1982). Nuevas consideraciones sobre la mejora genética de la prolificidad en el cerdo. In *2nd World Congress on Genetics Applied to Livestock Production*, Vol. 5, 548-67. Editorial Garsi, Madrid.

——, Martinat-Botté, F., Locatelli, A., Gruand, J., Terqui, M., and Berthelot, F. (1986). Components of prolificacy in hyperprolific Large White sows compared with the Meishan and Large White breeds. *Génét. Sél. Evol.* **18** (3) (in press).

Caritez, J. C. and Legault, C. (1985). Unpublished results.

Cheng, P. L. (1983). A highly prolific pig breed of China. The Taihu pig. *Pig News and Information* **4**, 407-16.

—— (1984). A highly prolific pig breed of China. The Taihu pig. (Parts III and IV). *Pig News and Information* **5**, 13-8.

Hill, W. G. and Webb, A. J. (1982). Genetics of reproduction in the pig. In *Control of pig reproduction* (ed. Cole, D. J. A. and Foxcroft, G. R.), 541-64. Butterworths, London.

Legault, C. (1985). Selection of breeds, strains and individual pigs for prolificacy. *J. Reprod. Fert.*, suppl. **33**, 151-66.

—— and Caritez, J. C. (1983). L'expérimentation sur le porc chinois en France. I. Performances de reproduction en race pure et en croisement. *Génét. Sél. Evol.* **15**, 225—40.

——, ——, Gruand, J., and Bidanel, J. P. (1984). Le point de l'expérimentation sur les races chinoises en France: reproduction et production. In *16èmes Journées de la Recherche Porcine en France*, pp. 481-94. Institut Technique du Porc, Paris.

—— and Gruand, J. (1976). Amélioration de la prolificité des truies par la création d'une lignée 'hyperprolifique' et l'usage de l'insémination artificielle: principe et résultats expérimentaux préliminaires. In *8èmes Journées de la Recherche Porcine en France*, pp. 201-8. Institut Technique du Porc, Paris.

——, Sellier, P., Caritez, J. C., Dando, P., and Gruand, J. (1985). L'expérimentation sur le porc chinois en France. II. Performances de production en croisement avec les races européennes. *Génét. Sél. Evol.* **17**, 133-52.

Roberts, R. C. and Smith, C. (1982). Genes with large effects. Theoretical aspects in livestock breeding. In *2nd World Congress on Genetics Applied to Livestock Production*, Vol. 6, 420-38. Editorial Garsi, Madrid.

Rombauts, P., Mazzari, G., and du Mesnil du Buisson, F. (1982). Premier bilan de l'expérimentation sur le porc chinois en France. 2. Estimation de composantes de la prolificité: taux d'ovulation et survie foetale. In *14èmes Journées de la Recherche Porcine en France*, pp. 137-42. Institut Technique du Porc, Paris.

Touraille, C., Monin, G., and Legault, C. (1983). Qualités organoleptiques des viandes de porcs croisés Piétrain/Chinois. In *15èmes Journées de la Recherche Porcine en France*, pp. 215-8. Institut Technique du Porc, Paris.

——, ——, and —— (1985). Qualités organoleptiques des viandes de porcs croisés Large White × Chinois. In *Proceedings of the 31st European Meeting of Meat Research Workers* (Varna, Bulgaria), vol. 2, pp. 790-3.

Wu, J.S. and Zhang, W.C. (1982). Genetic analysis of some Chinese breeds as a resource for world hog improvement. In *2nd World Congress on Genetics Applied to Livestock Production*, vol. 8, pp. 593-600. Editorial Garsi, Madrid.

Zhang, W.C., Wu, J.S., and Rempel, W.E. (1983). Some performance characteristics of prolific breeds of pigs in China. *Livest. Prod. Sci.* **10**, 59-68.

Discussion

It was suggested that the results with the pure-bred Chinese pigs compared with their crosses might be due to inbreeding. Seller said that this might be so particularly in respect of the Jiaxing breed, but in addition this breed differed in reaching its highest levels of prolificacy later in life. There was also a confusing effect in that many sows of this breed appeared to come into oestrus around 3 weeks after farrowing, but were not mated until after weaning at 4 weeks of age. The characteristics of the next heat appeared to be poor and this might confuse results.

There was discussion of the embryonic mortality observed in the Chinese pigs. It was pointed out that the results from China suggested that there might be higher ovulation rates than those observed in France and therefore that embryonic mortality would not differ very much. This appeared to be a possibility, but the general picture was confused because of age differences and the fact that older sows seem to give the highest ovulation rates.

18

Faster genetic improvement in sheep by multiple ovulation and embryo transfer (MOET)

C. Smith

Abstract

Faster rates of genetic change in sheep can be obtained by multiple ovulation and embryo transfer (MOET). The gain would come from increasing female reproductive rate, allowing shorter generation intervals or greater selection among females, or both. However, good embryo transfer rates are required at 6–8 months of age which are not achieved at present. With moderate rates of MOET (5 progeny per donor) gains in response of 50–70 per cent could be obtained for most traits, and with high MOET rates (10 progeny per donor) the gains in genetic response could be over 100 per cent.

Introduction

The value of multiple ovulation and embryo transfer (MOET) in increasing the rate of genetic change possible by selection has been shown by Land and Hill (1975) for beef cattle, and by Van Vleck (1981), Nicholas and Smith (1983), and Colleau (1985) for dairy cattle. In pig breeding the value of MOET is likely to be small (Smith 1981) because of the high reproductive rate and the short generation interval possible with pigs. Here an analysis of the value of MOET in sheep breeding (Smith 1986) is summarized.

Methods

The necessary techniques of MOET in sheep are available (e.g. Moore 1982) though at an experimental rather than a commercial level. However, survival of embryos from young females (6–8 months) is usually much lower than from adults (Quirke and Hanrahan 1977). This would have to be remedied to get the faster genetic responses indicated here, since they are based on MOET of young females. The gains come from reducing the generation interval, or from increased selection among females, or from both.

Sheep breeds are variable in age at first breeding, and in litter size and intermediate values are used; the number of progeny available for selection per

163

female mated is taken as 0.5, 1.0, and 1.5 for 1 year, 2 year, and older females, respectively. Mating ratios (females per male) of 25, 50, and 100 are studied. For MOET, rates of 2, 5, and 10 progeny (per donor per breeding season) are examined, and the mating ratios are reduced to 5, 10 and 20, to maintain moderate levels of inbreeding.

Three groups of traits are considered: (1) Traits measured on the live animal before reproduction, such as growth and (*in vivo*) carcass traits; (2) wool traits measured at 14–16 months of age; and (3) female reproductive traits. Generation intervals possible for the three groups of traits by normal reproduction, MOET and progeny testing of males are given in Table 18.1, along with the range in heritabilities. Standard methods (e.g. Falconer 1981) were used to estimate the annual rates of genetic change for the different traits and breeding systems.

Results

Growth and carcass traits

For growth and carcass traits measured before reproductive age, MOET allows the minimum generation interval to be reduced from 1.5 to 1.0 years, and allows some selection among females. With these changes much higher genetic responses from individual (mass) selection can then be obtained by using MOET; some 60–80 per cent higher with moderate MOET rates, and up to 100 per cent higher for high MOET rates, as shown in Fig. 18.1. For traits

Table 18.1. Generation intervals with normal reproduction (NR), embryo transfer (MOET) and progeny testing of males (PT), and heritabilities for different traits in sheep

Trait	Selection method	Breeding ages (years)		Heritability
		Males	Females	
Growth and carcass traits	NR	1	1,2,3	0.1–0.2*
	MOET	1	1	0.2–0.4†
	PT	2	1,2,3	
Wool yield	NR	2	1,2,3	0.3–0.4
	MOET	2	2	
	MOET‡	1	1	
Litter size	NR	1	1,2,3	0.05–0.15
	MOET	1	1,2,3	
	PT	3	1,2,3	

* Growth.
† Leanness.
‡ Juvenile MOET, selection at 18 months.

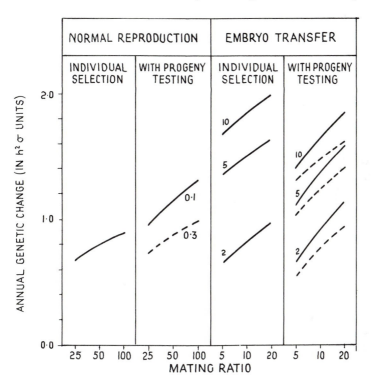

Fig. 18.1. Estimated rates of annual genetic change (in units of $h^2\sigma$) in traits measured before breeding age, with normal reproduction and with MOET. Rates are shown for individual selection alone, and for individual selection followed by progeny testing of males (with two levels of heritability, 0.1 and 0.3). MOET rates are 2, 5, and 10 progeny per donor.

with low heritability progeny testing of males might be used, in addition to individual selection and embryo transfer. However, as also shown in Fig. 18.1, progeny testing does not further enhance the genetic responses.

Wool yield

Selection for wool yield is usually at 14–16 months of age, with most of the selection on males. With MOET, it would be possible to reduce the generation interval in females to 1 year (without selection), or to select among females at 2 years of age. The extra genetic responses possible with these alternatives are shown in Fig. 18.2, ranging from 30–40 per cent, for moderate MOET rates, and up to 60 per cent for high MOET rates. A novel alternative allows even higher rates of genetic response. This juvenile MOET scheme (after Nicholas and Smith 1983) involves mating between unselected males and females at 6–7 months of age and selecting progeny from the best matings, as assessed by parental records at 14–16 months of age. The juvenile scheme allows rates of genetic response to be more than doubled, as shown in Fig. 18.2.

Litter size

With its low heritability, the best genetic responses with natural reproduction come from selection on a family index at 6–7 months of age (Martin and Smith 1980). Since progeny testing of males takes 3–4 years, it is not competitive. To use MOET effectively requires the females to be bred for MOET (early in the season) and then for normal reproduction, so as to obtain a litter record. With MOET and larger family sizes, selection can be more accurate and more intense for females. These systems allow the rates of genetic change to be increased by 50–100 per cent, compared with normal reproduction (Fig. 18.3), and show possible annual rates of genetic change over 5 per cent of the mean per year — a remarkable rate for a trait of low heritability (0.1) expressed in only one sex after reproductive age.

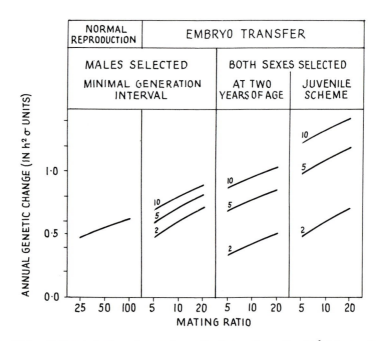

Fig. 18.2. Estimated rates of annual genetic change (in units of $h^2\sigma$) in wool yield, with normal reproduction and with MOET at 2 years of age, or at 1 year of age (juvenile MOET scheme involving mating and transfer before selection). Other details as in Fig. 18.1.

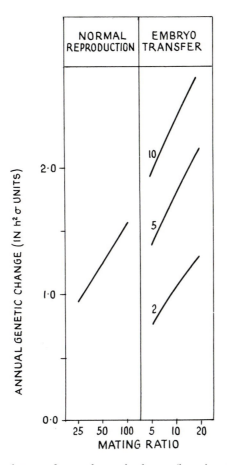

Fig. 18.3. Estimated rates of annual genetic change (in units of $h^2\sigma$) in litter size, with normal reproduction and with MOET (followed by a normal litter record in the same year). Other details as in Fig. 18.1.

Discussion

The results show that use of MOET can allow substantial increases in the rates of genetic response possible for different traits in sheep breeding. They are, however, dependent on MOET at an early age (6–8 months), and research is needed to improve the yield and survival of embryos from young donor females. Selection will usually be for an index of several economic traits, often measured at different ages so requiring sequential selection schemes. With the MOET systems, selection is largely at 6–8 months of age, either on individual performance or on family index, so the problem of differing ages for different traits is avoided. The increases in response by

MOET could be supplemented by other means, for example by using any physiological indicators of merit (Walkley and Smith 1980), by embryo splitting, and by earlier breeding using hormone or light pattern treatments. With MOET, and faster generation turnover, the rates of inbreeding will be increased, and larger breeding populations will be needed to moderate the inbreeding rates. This, together with the expertise needed for MOET, index selection, and to run an efficient breeding programme, suggests that the systems would be best carried out in a large specialized nucleus breeding unit, rather than by dispersed farmer-breeders. The national benefits from the genetic improvements possible by MOET in sheep, are large (Smith 1978) and should be exploited.

References

Colleau, J. J. (1985). Genetic efficacy of embryo transfer in nucleus selection herds of dairy cattle. *Genet. Select. Evol.* **17**, 499–538.

Falconer, D. S. (1981). *Introduction to Quantitative Genetics* (2nd edn.). Longman. London.

Land, R. B. and Hill, W. G. (1975). The possible use of superovulation and embryo transfer in cattle to increase response to selection. *Anim. Prod.* **21**, 1–12.

Martin, T. G. and Smith, C. (1980). Studies on a selection index for improvement of litter weight in sheep. *Anim. Prod.* **31**, 81–5.

Moore, N. W. (1982). Egg transfer in the sheep and goat. In *Mammalian Egg Transfer* (ed. C. E. Adams) pp. 119–33. CRC Press, Florida.

Nicholas, F. W. and Smith, C. (1983). Increased rates of genetic change in dairy cattle by embryo transfer and splitting. *Anim. Prod.* **36**, 341–53.

Quirke, J. F. and Hanrahan, J. P. (1977). Comparison of the survival in the uteri of adult ewes of cleaved ova from adult ewes and ewe lambs. *J. Reprod. Fert.* **51**, 487–9.

Smith, C. (1978). The effect of inflation and form of investment on the estimated value of genetic improvement of farm livestock. *Anim. Prod.* **26**, 101–10.

—— (1981). Levels of investment in testing and in the genetic improvement of livestock. *Livest. Prod. Sci.* **8**, 193–202.

—— (1986). Use of embryo transfer in genetic improvement sheep. *Anim. Prod.* **42**, 81–88.

Van Vleck. L. D. (1981). Potential genetic impact of artificial insemination, sex selection, embryo transfer, cloning and selfing in dairy cattle. In *New technologies in animal breeding* (eds. B. G. Brackett, G. E. Seidel Jr, and S. M. Seidel), pp. 221–242. Academic Press, London.

Walkley, J. R. W. and Smith, C. (1980). The use of physiological traits in genetic selection for litter size in sheep. *J. Reprod. Fert.* **59**, 83–8.

Discussion

Hanrahan suggested that when dealing with litter size in sheep it might be possible to double the rate of progress by selecting on ovulation rate. This was thought to be useful and add further to a MOET scheme, and something which could be incorporated once the appropriate evidence for its efficacy was produced. The problem of embryonic mortality experienced in eggs from young ewes was further considered. Hanrahan explained that in two breeds, the Galway and the Romney, this problem had been found, but had not been observed in the Romanov. It might therefore merely be a matter of maturity and that by waiting a further 3 months, the problem could possibly be avoided.

The use of AI in sheep in Britain was raised and it was explained that practically none was practiced. It was therefore felt that there might be some balance between extending the use of artificial insemination and introducing embryo transfer. In reply Smith pointed out that progeny testing was not of great utility in the breeding programmes studied, but AI could be used to increase mating ratios and speed up dissemination of improvement. The utility of frozen sheep embryos for export was discussed, this having recently been carried out in Denmark for the transport of stock to New Zealand.

The question of using an open nucleus was raised and this was agreed to be sensible using better animals wherever they could be found, at home or abroad.

19

The use of marker loci in selection for quantitative characters

P. Stam

Abstract

A theoretical framework is offered through which the merits of marker assisted sib selection, relative to normal sib selection of young bulls in a dairy cattle breeding programme can be evaluated. It is shown that in the theoretical ideal circumstances for marker-assisted sib selection, the expected gain in selection response is approximately 40 per cent. Imperfections caused by recombination reduce this extra selection response. Since the full application of marker assisted selection by means of restriction fragment length polymorphisms, to a breeding programme implies a considerable financial investment, a rough calculation was made of the risk that marker assisted selection will, in any generation, do worse than normal sib selection. Once a number of markers are available for this purpose, a simple analysis of variance may shed some light on the prospects of marker assisted sib selection.

Introduction

The advent of modern molecular techniques that enable the detection of restriction fragment length polymorphisms (RFLPs), has raised renewed interest in the use of genetic marker loci as an aid in estimating and/or predicting genotypic values or breeding values (BV) (Soller 1978; Soller and Beckman 1982; Roberts and Smith 1982). The reason for this revival obviously is the large number of RFLPs that is, in principle, available for this purpose. It has been suggested (Soller and Beckman 1982) that, by means of RFLPs, the genome of an organism could be marked so heavily that knowledge of the complete marker genotype of an individual enables a fairly precise prediction of its breeding value with respect to a quantitative trait.

Selection based on such a predicted BV is called marker assisted selection (MAS). It could be used in a breeding programme for dairy cattle, e.g. by pre-selection of young bulls (that have not yet been progeny-tested) on the basis of their marker genotype.

The purpose of the present paper is to offer, by using an elementary quantitative genetic approach, some insight into the prospects of MAS.

Attention is also paid to its merits relative to the classical selection procedures.

Segregating marker alleles can, of course, only be useful for selection purposes, when they are associated with the alleles of the loci involved in the quantitative trait of interest. Association is complete when a (polymorphic) restriction site is located within the coding sequence of a 'quantitative' gene, though this in itself is no guarantee that the alternative alleles have any effect on the trait. In most cases, however, associations of marker and 'quantitative' alleles will arise through linkage. Fundamental population genetics theory tells us that at the population level associations of the alleles of any pair of loci, i.e. linkage disequilibrium, may not be expected in a large random mating population, unless both loci are under selection pressure in a specific way. (Apart from accidental linkage disequilibrium as a result of random drift, recent admixture of gene pools from distinct origin may result in considerable linkage disequilibrium.) In a large population, association of alleles is to be expected within the offspring of a single double heterozygous individual. As a simple example, which illustrates the principle for detection of associations of marker alleles and quantitative alleles, consider the random (half-sib) offspring of an $A_1A_2B_1B_2$ individual, where A and B denote the marker and quantitative locus, respectively (see Table 19.1). In Table 19.1 c denotes the recombination frequency, A. denotes an undefined allele of the A-locus.

Table 19.1.

Parent	Gametes	Frequency	Offspring	Mean phenotypic value
	A_1B_1	$(1-c)/2$	$A_1A.B_1B.$	
	A_1B_2	$c/2$	$A_1A.B_2B.$	$\bar{x}_1$
A_1B_1				
A_2B_2	A_2B_1	$c/2$	$A_2A.B_1B.$	
	A_2B_2	$(1-c)/2$	$A_2A.B_2B.$	$\bar{x}_2$

Suppose that the offspring can be classified with respect to the marker allele they received from the common parent. Then, the difference between the mean phenotypic values ($\bar{x}_1$ and $\bar{x}_2$) of these groups is, to a certain extent, a measure of the effects of the alleles at the B-locus. Of course, the value of c, or the degree of association will, influence the measured difference $\bar{x}_1 - \bar{x}_2$. It is important to realize that associations of this kind can only be detected within the half sib offspring of a single individual. This is because in the HS offspring of a second individual, not only may the associations be reversed, but also other alleles of both the marker and the quantitative locus may be involved. It will be clear that reliable estimates of allele-effects, as measured by the difference $\bar{x}_1 - \bar{x}_2$, can only be obtained when a single individual has a fairly large number of HS-offspring. For this reason, the following analysis was done while having in mind a dairy cattle breeding programme with large scale AI.

Analysis

From the point of view of marker aided selection, the 'best' situation one can think of is one with complete association of marker and quantitative alleles. This is the case when the marker genes are in fact the same ones as the polygenes involved in the trait of interest. In no other situation can a larger proportion of the variation within a HS family be 'explained' by qualitative differences. For this reason we will start the analysis by considering a single multi-allelic marker locus which (also) is responsible for the total quantitative genetic variation. A single locus model for the control of a quantitative character, with qualitatively identifiable alleles, may seem highly unrealistic. For the present purpose, however, it is useful in the sense that it represents the perfect circumstances for marker-assisted selection. Any imperfection, such as polygenic control with incomplete linkage, will cause MAS to be less effective. Thus, the simple model will indicate the upper limits of the gains that are to be expected from MAS.

The model is specified by a series of indentifiable alleles, $A_1, A_2, .., A_n$. The average effects of the alleles are denoted by $\alpha_1, .., \alpha_n$. For simplicity dominance will not be considered. So the genotypic value (measured as deviation from the population mean) of the genotype $A_i A_j$ equals $\alpha_i + \alpha_j$. Table 19.2 outlines the procedure of MAS of young bulls whose female sibs have been classified and measured. These sibs constitute a HS family sired by a

Table 19.2. Simple prediction of a young bull's genotypic value by means of classified and measured (female) half sibs

Production bull	Paternal gametes	Maternal gametes	Mean phenotypic value of classified daughters
		$A_1........A_n$	
	A_1	classified	$\bar{x}_1$
A_1A_2			
	A_2	daughters	$\bar{x}_2$
Young bull	Mean genotypic value (y)		Predicted genotypic value ($\hat{y}$)
$A_1A.$	α_1		$\bar{x}_1$
$A_2A.$	α_2		$\bar{x}_2$

heterozygous production bull (A_1A_2). The classification of sibs is according to which of the paternal alleles was passed to them.

Young bulls sired by the same production bull are classified with respect to their paternal allele. Their genotypic values are predicted by the mean phenotypic value of their sisters (measured as deviation from the population mean) which received the same paternal allele. This scheme follows the procedure sketched by Soller and Beckman (1982). It is obvious that both sons and daughters in the sibship which have the same genotype as the common parent cannot be classified with respect to their paternal allele. Furthermore, the 'genotyping' of a complete female sibship could be too costly in practice. Therefore, we shall slightly extend the procedure as follows. We assume that a fraction p of the female offspring has been classified with respect to their paternal allele and that the remaining fraction ($q = 1 - p$) is unclassified (or unclassifiable). We assume further that measurements of the trait are available for both classified and unclassified daughters. The corresponding prediction procedure is outlined in Table 19.3. The selection criteria for young bulls is their predicted genotypic value (y). The relative amounts of information provided by classified and unclassified sibs enter as weight factors in the predictors. When $p = 0$, i.e. no genotyping of sibs, the predicted value for all young bulls in a sibship is the same ($\hat{y}_1 = \hat{y}_2 = \hat{y}_u = \bar{x}_u = \bar{x}$), and the selection criterion corresponds to normal half-sib selection. When $p = 1$ we have what might be called full marker assisted sib selection. In order to obtain an impression of the merits of MAS, relative to normal HS selection, we have to calculate the correlation coefficient, $r_p(y,\hat{y})$, between true and predicted genotypic values of young bulls, using $p = 0$ as a yardstick. Referring to the usual equation for selection response,

$$R = i\,\sigma_y\,r(\hat{y},y),$$

we see that the ratio of selection responses equals

$$R_{MAS}/R_{HSS} = r_p(\hat{y},y)/r_0(\hat{y},y),$$

where the subscript (p,0) refers to proportion of classified sibs. We will express the extra selection response due to MAS as the gain

$$\begin{aligned} G &= (R_{MAS}/R_{HSS}) - 1 \\ &= \{r_p(\hat{y},y)/r_0(\hat{y},y)\} - 1. \end{aligned}$$

With the usual definitions of gene effects (α_i) it is straightforward to calculate var(y), var($\hat{y}$) and cov($\hat{y},y$). One arrives at

$$\text{var}(y) = V_A, \tag{1a}$$

$$\text{var}(\hat{y}) = \frac{1+p}{n}\left(V_E + \frac{1}{2}\,V_A\right) + (1+p^2)\frac{1}{4}V_A + \frac{1-p}{n}\,\frac{1}{4}V_A, \text{ and} \tag{1b}$$

$$\text{cov}(\hat{y},y) = (1+p)\frac{1}{4}V_A, \tag{1c}$$

where n is the total size of a female sibship, V_A is the additive genetic variance ($V_A = 2\varepsilon\alpha^2$), V_E is the environmental variance.

Table 19.3. Prediction of a young bull's genotypic value when not all of its female sibs have been classified. $\bar{x}_u$ denotes the mean phenotypic value of unclassified sibs; x denotes the general mean of the female sibship

		A_1........A_n	Class means
$\dot{A}_1A_2$	$\frac{1}{2}p$ A_1	classified	$\bar{x}_1$
	$\frac{1}{2}p$ A_2		$\bar{x}_2$
	$1-p$ A.	unclassified	$\bar{x}_u$

Young bull	Mean genotypic value (y)	Predicted genotypic value ($\hat{y}$)
$A_1A.$	α_1	$\hat{y} = p\bar{x}_1 + q\bar{x}_u$
$A_2A.$	α_2	$\hat{y} = p\bar{x}_2 + q\bar{x}_u$
$A.A.$	$\frac{1}{2}(\alpha_1 + \alpha_2)$	$\hat{y}_u = \bar{x}$

In the derivation of equation (1b), it has been assumed that the segregation ratio of marker alleles among classified sibs is exactly 1:1. Variation in the realized segregation ratio's introduces an extra term of the order $1/n$ in the expression for var($\hat{y}$). The meaning of equation (1b) can best be seen by taking $p=0$ and $p=1$. For $p=0$ we have

$$\text{var}(\hat{y}) = \frac{1}{4}V_A + \frac{1}{n}\left(\frac{3}{4}V_A + V_E\right), \tag{2}$$

the two familiar terms, corresponding to the variance of HS family means and mean variance within HS families, respectively.

For $p=1$ we have

$$\text{var}(\hat{y}) = \frac{1}{2}V_A + \frac{1}{\frac{1}{2}n}\left(\frac{1}{2}V_A + V_E\right). \tag{3}$$

Now, the two terms correspond to the variance between and within classified daughter groups. For $n=\infty$ and $p=1$, var(y) measures the variance of allele effects, var(α) which equals $\frac{1}{2}V$.

Using equations (1a–1c), the correlation coefficient becomes

$$r_p(\hat{y},y) = \frac{(1+p)\ \dfrac{1}{4}\ V_A}{\sqrt{\left(\dfrac{1+p}{n}\left(\dfrac{1}{2}\ V_A + V_E\right) + (1+p^2)\ \dfrac{1}{4}\ V_A + \dfrac{1-p}{n}\ \dfrac{1}{4}\ V_A\right)}}$$

which can be rewritten as

$$r_p(\hat{y},y) = \frac{\dfrac{1}{4}\ (1+p)}{\sqrt{\left(\dfrac{1+p}{n}\ \dfrac{2-h^2}{2h^2} + \dfrac{1+p^2}{4} + \dfrac{1-p}{4n}\right)}} \tag{4}$$

where $h^2 = V_A/(V_A + V_E)$.

As an illustration of equation (4), let $n = \infty$ and $p = 0$ or $p = 1$. For $p = 0$ (i.e. HSS), $r = 0.5$, which means that 25 per cent of the genetic variation among young bulls is 'explained' by the predictor. For $p = 1$ (MAS), we have $r\sqrt{0.5}$, so that now 50 per cent of variation among young bulls is explained by the predictor. The gain in selection response due to MAS thus equals $2\sqrt{0.5} - 1$, or approximately 40 per cent. Since we are dealing with the most favourable model for MAS, this figure represents an upper limit.

In order to check the validity of equation (4), a simulation of MAS, using a single multi-allelic locus, was carried out. A flow diagram of the simulation program is given in Fig. 19.1. Neither the number of segregating alleles, nor the magnitude of the allele effects has an effect on the simulation results. All simulation results presented were done with at least 20 replicate runs and 100 breeding sires per replicate. Per sire at least 10 sons were generated.

The theoretical relation of equation (4), together with a sample of simulation results are given in Figs. 19.2–19.4. It can be seen that there is a close agreement between theoretical values and the ones obtained by simulation. Only for family sizes less than 10 does the variation in actual segregation ratio's reduce the correlation slightly below its theoretical value. The effect of p, the proportion of classified sibs on $r(y,\hat{y})$ is seen to be approximately linear (Fig. 19.4). So classifying, e.g. only 50 per cent of the total number of sibs implies a loss of approximately 50 per cent of the potential gain in selection response. On the other hand, keeping $n.p$ fixed in equation (4) shows, that once a reasonable number of sibs have been classified (say 20 or more), the addition of unclassified sibs hardly increases the correlation.

Multiple loci and recombination

When leaving the simple one-locus model, we have to consider the effect of recombination between markers and quantitative loci. The effect of

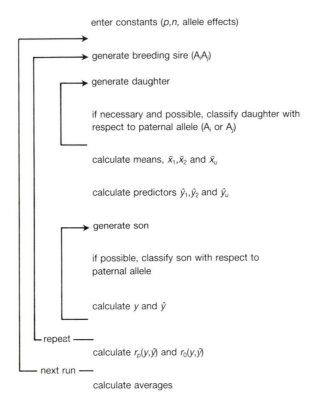

enter constants (p,n, allele effects)

generate breeding sire (A_iA_j)

generate daughter

if necessary and possible, classify daughter with respect to paternal allele (A_i or A_j)

calculate means, $\bar{x}_1, \bar{x}_2$ and $\bar{x}_u$

calculate predictors $\hat{y}_1, \hat{y}_2$ and $\hat{y}_u$

generate son

if possible, classify son with respect to paternal allele

calculate y and $\hat{y}$

repeat ——

calculate $r_p(y,\hat{y})$ and $r_0(y,\hat{y})$

next run ——

calculate averages

Fig. 19.1.   Flow sheet of simulation program for MAS.

recombination on the correlation $r(\hat{y},y)$ can however be studied with the one-locus model. This is because the result of a recombination event during meiosis in the breeding sire simply results in a misclassification of a daughter. Using the symbol c for the frequency of misclassification the correlation coefficient for large completely classified sibships (i.e. by neglecting terms of the order $1/n$ in the relevant expression for var(y)), one finds

$$r_c(\hat{y},y) = \sqrt{[0.5\{1 - 2c(1-c)\}]} \tag{5}$$

Taking $c = 0$ we find $r = \sqrt{0.5}$, as it should; with $c = 0.5$ we have free recombination which means that 50 per cent of the sibs are misclassified. In that case $r = 0.5$, the same value as if only unclassified daughters has been available. Equation (5) suggests the relation

$$r_c = r_{c=0.5}\sqrt{[2\{1 - 2c(1-c)\}]} \tag{6}$$

where $r_{c=0.5}$ stands for the correlation in case of free recombination. In order to check whether equation (6) also holds in case of finite sibships, an adapted

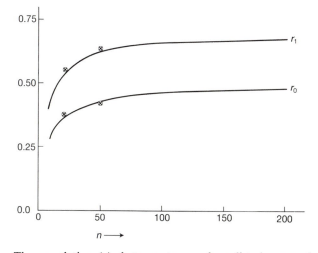

Fig. 19.2. The correlation (r), between true and predicted genotypic values of young bulls for different sibships sizes (n). r_0: HSS; r_1: MAS; solid line: equation (4); x: average of 20 simulation runs.

version of the simulation program (allowing recombination) was run for several combinations of parameters (n and h^2). It turned out that equation (6) describes the effect of recombination satisfactorily for $c < 0.40$ and $n > 40$. Figure 19.5 examplifies the effect of recombination. The general trend, also illustrated by Fig. 19.5, is a rather drastic effect of recombination; e.g. with 15 per cent recombination approximately 70 per cent of the gain in selection response that could be obtained by MAS if there were no recombination, is left.

Extension of the theory to a model with quantitative polygenes, dispersed over the genome, in combination with multiple markers, is rather difficult. However, the present approach shows that the crucial point for MAS to be successful is the proportion of quantitative variation among gametes that is associated with qualitative variation. When multiple markers are available, predictors for young bull's BVs can be obtained by means of multiple regression (within sibships) of the trait value of classified daughters on the 'independent' variables; the latter indicate, as before, which paternal marker allele was passed. (The predictors for the one-locus model with $p = 1$ are merely a special case of multiple regression.) It is tempting to guess what proportion of the variation between paternal gametes could possibly be 'explained' by a large number of markers. If markers were available at an average cross-over distance of 40 centimorgan, and the polygenes for the trait of interest were more or less uniformly distributed over the genome, one might take $c = 0.15$ as the average recombination frequency between a

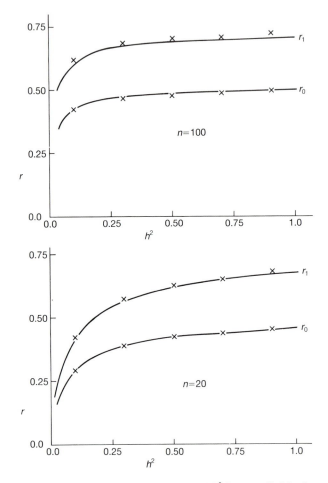

Fig. 19.3.   The correlation $r(\hat{y},y)$ as a function of h^2 for two sibship sizes, $n = 20$ and $n = 100$. r_0: HSS; r_1: MAS. Further legend as in Fig. 19.2.

marker and its nearest quantitative gene. Equation (6) tells us that the gain in selection response due to MAS will in that case be approximately 20 per cent. Compared with the theoretical maximum gain of 41 per cent this figure is rather disappointing.

Discussion

Large scale application of MAS, exploiting RFLPs, implies a considerable financial investment. From an investor's point of view it is not only

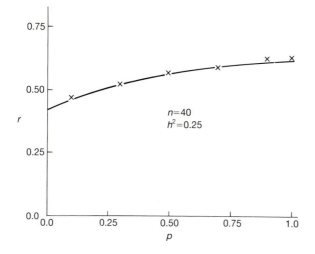

Fig. 19.4. The correlation $r(y,\hat{y})$ as a function of p, the proportion of classified sibs. Further legend as in Fig. 19.2.

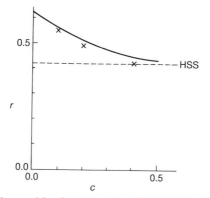

Fig. 19.5. Effect of recombination (c) on the $r(\hat{y},y)$. HSS refers to ordinary half-sib selection.

important to know what the expected pay-off will be, but the risk should also be known. It is, of course, rather difficult to predict financial returns from expected genetic improvement. However, one might consider the following question. How sure can one be about the additional genetic improvement by applying MAS instead of the classical selection procedures? To answer this question in detail one should know the probability distribution of the gain in selection response in a real breeding programme. We are in a position to get a quick impression of this distribution by means of simulation, using assumed values of $r_0(\hat{y},y)$ and $r_1(\hat{y},y)$, corresponding to HSS and MAS, respectively. The scheme for such a simulation program is as follows.

1. Generate a number of normal random deviates, representing the true genotypic values of young bulls.
2. Associated with each genotypic value, generate two predictors, also normally distributed, according to the assumed numerical values of the correlations, r_0 and r_1.
3. Rank the items with respect to the two selection criteria, i.e. the values of the predictors.
4. Select, using each of the criteria, a certain fraction of the 'best' items.
5. Calculate, for either of the two selection criteria, the mean genotypic value of the selected group.
6. Calculate the realized gain in 'response' obtained by using the better of the two predictors, i.e. the one with the highest correlation.
7. Repeat steps 1–6 a large number of times and produce a frequency distribution of the gain in response.

Figure 19.6 shows a sample of results obtained in this way, based on 500 replicate runs. It is seen that the realized gain in selection response is quite

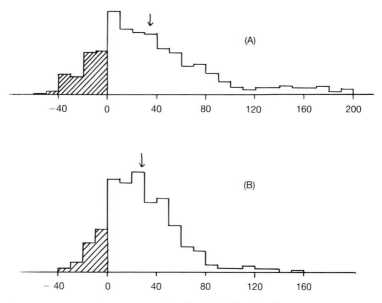

Fig. 19.6.   Frequency distribution of realized gain in selection response (expressed as a percentage) obtained by using a superior predictor. $r_0 = 0.43$; $r_1 = 0.53$; (A) 20 items selected from a total of 200. (B) 100 items selected from a total of 200. Hatched areas indicate the probability of obtaining less response when using the superior predictor. Arrows indicate the means of the distributions.

variable. In this particular case the range of the gain is from -60 per cent to $+200$ per cent. The probability of obtaining less response with the superior selection criterion is approximately 15 per cent. Of course, these results apply to a single cycle of selection only; in the long run, the superior predictor will almost certainly outdo the inferior one. Nevertheless, the large variation in extra response per generation when using MAS is worth keeping in mind.

RFLPs have often been considered useful in detecting 'major genes', or quantitative alleles with large effects (see e.g. Soller *et al.* 1976; Roberts and Smith 1982). The problems in identifying genes with effects have been reviewed and discussed by Roberts and Smith (1982). The approach of the present paper is not in terms of major loci, but generalizes to 'polygenes' with small individual effects, the number of loci involved and the magnitude of the allele effects being irrelevant for this approach. As stated earlier, the main question with respect to the prospects of MAS is: what proportion of the genetic variance among gametes is likely to be associated with markers? Using normal RFLPs this proportion seems not to guarantee a trade-off. The recent discovery of hypervariable DNA regions (a variable copy number of tandem repeats in these regions) that occur at some 20 sites in the human genome, is however rather promising in this respect (Davies 1985; Jeffreys *et al.* 1985). The potential power of this type of markers lies in the fact that multiple loci can be screened simultaneously by means of a single probe. If such hypervariable regions are also found in livestock, they may be useful for MAS. The question remains, however, what proportion of the genetic variance they will turn out to be associated with. The proof of the pudding is in the eating. Once a large number of markers are available for this purpose, a one generation multiple regression analysis will reveal which proportion of the variance among classified sibs is attributable to qualitative differences. The 'explained' variance may range from 0 to $\frac{1}{4}V_A$. The numerical results of this paper suggest a linear relationship between 'explained' variance and expected gain in selection response, with an upper limit of approximately 40 per cent.

References

Davies, K. E. (1985). New hypervariable DNA markers for mapping human genetic disease. *Trends Genet.* **1**, 97.

Jeffreys, A. J., Wilson, V., and Thein, S. L. (1985). Hypervariable 'mini-satellite' regions in human DNA. *Nature* **314**, 67–73.

Roberts, R. C. and Smith, C. (1982). Genes with large effects. Theoretical aspects in livestock breeding. *2nd World Congress on genetics applied to livestock production*. Madrid.

Soller, M. (1978). The use of loci associated with quantitative traits in dairy cattle improvement. *Anim. Prod.* **27**, 133–9.

—— and Beckman, J. S. (1982). Restriction fragment length polymorphisms and genetic improvement. *2nd World Congress on genetics applied to livestock production*. Madrid.

——, Genizi, A., and Brody, T. (1976). On the power of experimental designs for the detection of linkage between marker loci and quantitative loci in crosses between inbred lines. *Appl. Genet.* **47**, 35–9.

Discussion

Petersen pointed out that the derivations assumed complete linkage, i.e. pleiotropy, and Stam agreed. Hill asked if the estimates of the effects were all within sire groups. Stam agreed, but pointed out that they could be combined across families if half-sib groups are homogeneous. Smith pointed out that the maximum gain of 40 per cent is only possible if all sires are doubly heterozygous, but Stam suggested that his simulation studies indicate that the reduction due to segregation in a population is small. Land wondered if the analysis was restricted to half-sib groups, and whether it could be generalized. Stam felt that half-sib groups were the natural context in which to detect and study such associations, but full-sib groups could also be used. Smith asked if the possible 40 per cent gain in accuracy by screening young sires before progeny testing, meant less return from progeny testing. Stam agreed, but felt that the 40 per cent gain would not be achieved in practice on a large scale, and so the reduction in response from progeny testing would be small.

20
Single sex beef cattle systems

St. C. S. Taylor, R. B. Thiessen, and A. J. Moore

Abstract

With sexed embryo transfer or sexed sperm, new directions in animal production can be anticipated. In beef cattle, the likely impact is evaluated by comparing efficiency of food utilization in traditional and sex-controlled systems of production. A dam and all her progeny are taken as a production unit. Conditions for maximum efficiency, including optimal age at slaughter, are calculated for each system. Production systems are then compared for maximum efficiency. Traditional systems achieve their maximum when a dam produces on average just enough calves to provide her own replacement. The optimal slaughter point usually occurs when progeny have attained 0.5 to 0.7 of their mature weight.

The main result is that a single-sex bred-heifer (SSBH) system could be much more efficient (by up to 27 per cent) than any traditional system. Maternal overhead costs are largely eliminated. Surprisingly an 'all-male' system has not a great deal more to offer than a traditional system that uses a large terminal sire. An SSBH system with an 80 per cent reproductive rate would require 20 per cent of dams to have a second calf and is thus exceptionally suitable for practising genetic selection.

An SSBH system should become competitive with traditional systems when the costs of breeding by sexed sperm or sexed embryo transfer are less than ten times normal breeding costs.

Introduction

In most animal production systems, but especially in beef cattle, maternal overheads are a significant part of production costs. The food consumed by the dam relative to the food consumed by her progeny might be reduced in several ways. (1) Increasing litter size or producing twins in cattle increases output without increasing maternal overheads. (2) Using a large terminal sire does the same. (3) Genetic selection for dams with higher maintenance efficiency directly reduces maternal overheads. (4) Using bred heifers greatly reduces maternal overheads because the dam is herself still growing and therefore still productive. This paper is mainly concerned with evaluating and exploiting the advantages of bred heifers in the advent of sex control.

Briefly, a single-sex bred-heifer line leads to high food efficiency because the maternal overhead costs of producing a calf are greatly reduced. A bred heifer growing towards slaughter is like a maiden heifer with two additional costs. One is the cost of pregnancy which cannot be avoided; but it is only 8 per cent of the maternal overhead cost of maintaining a fully mature dam between calvings. The second difference is that in commercial practice the bred heifer is normally about 0.8 mature when slaughtered after recovery from calving, whereas the maiden heifer is slaughtered when about 0.6 mature. This delay represents a marked drop in the food efficiency achievable. Among breeds currently available, only a Jersey heifer could produce a calf when 0.6 to 0.7 mature, but she would not yield a high quality beef carcass. Producing beef with sex control thus leads to a new direction for genetic selection, namely, for an early breeding genotype with a high quality carcass when about 0.6 mature.

Materials and methods

The model for overall food efficiency

The AFRC Animal Breeding Research Organization's multibreed cattle experiment (Thiessen *et al.* 1984) has now provided basic information necessary for the construction of a mathematical model of efficiency of food utilization in different systems of beef production.

The model describes the efficiency of conversion of food into lean meat, the primary end product of beef production. Meat is produced by slaughtered offspring (male and female) and by the dam when culled. Food is consumed by slaughter offspring from birth to slaughter and by the dam from birth until culling after a specified number of calvings. Overall food efficiency is therefore defined as total lean meat produced divided by total food intake in a production unit consisting of a dam and all her progeny except for one female replacement calf which is excluded. All tables and figures are based on these overall food efficiencies.

In the model, parameter values specify production conditions in terms of number of calvings per dam, size of terminal sire, maintenance cost or efficiency of dam, reproductive rate, and breeding cost with or without sex control; also degree of maturity in body weight (u) defined as body weight expressed as a proportion of adult body weight. A detailed description of the mathematical construction of the model and its properties is given by Taylor *et al.* (1985).

In the present context, a system operating under 'best' conditions means a reproductive rate (r) of 0.85, with the cost of a unit of food fed to the dam being 0.6 times that of food fed to progeny, and bred heifers slaughtered when 0.6 mature in body weight; all compared with 'normal' conditions where reproductive rate is 0.75, unit food cost is the same for dam and progeny, and bred heifers are slaughtered when 0.8 mature. A large terminal sire breed is assumed to give progeny 1.4 times heavier than a standard pure-bred sire. The number of calvings per dam needed to obtain a female replacement is $2/r$.

Results and discussion

Maximum efficiency and optimal degree of maturity at slaughter

For a production system with a given set of conditions (parameter values), overall food efficiency always had a maximum value which defined the optimal degree of maturity for slaughter of progeny. Maximum overall food efficiency and optimal degree of maturity at slaughter were linearly related as represented by the straight line in Figs. 20.1, 20.5 and 20.6. A major result, which can be seen from those figures, was that the change in overall efficiency as degree of maturity at slaughter (u_s) departed from its optimum value, was surprisingly small in the range $u_s = 0.4$–0.7. Other criteria such as carcass quality, market price, food cost, and availability can therefore be allowed to determine when progeny are slaughtered without incurring a severe food-efficiency penalty.

Decline in maximum overall food efficiency with number of calvings per dam

Another major result obtained from the model was that in a pure-bred beef production system and in all but the most efficient cross-breeding systems, overall efficiency declined as number of calvings per dam increased. In

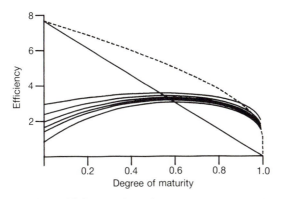

Fig. 20.1. Overall food efficiency (g lean tissue per MJ ME) as a function of degree of maturity of progeny at slaughter in a traditional purebred system under best conditions with number of calvings per dam increasing from $2/r$ (top curve) to 3, 4, 5, 6, and 10 (bottom curve). Also straight line giving points of maximum efficiency and curve (--) showing upper limit to efficiency when there are no maternal overheads. (Reproduced by courtesy of Animal Production.)

general, therefore, the sooner a replacement female was obtained and the dam slaughtered, the higher was the efficiency of the system. This may mean that in times of food shortage, older females should have a lower priority for survival if efficiency of food utilization is important.

Curves of overall efficiency in relation to degree or maturity of progeny at slaughter are given in Fig. 20.1 for different numbers of calvings per dam in a traditional purebred system. The decline in overall efficiency is typical of many systems, three of which are now examined.

Traditional system

The decline in efficiency with number of calvings per dam was quite steep for a pure-bred system but was much reduced when a large terminal sire was used (Fig. 20.2). The magnitude of the decline for a normal pure-bred sire and a large terminal sire under different conditions is given in Table 20.1. Unless the sire breed was about twice as large as the dam breed, and even then only under best conditions, there was a decline in efficiency with increasing age of the dam at culling. In theory, crossing to a sire breed larger than the dam breed always increased food efficiency, although in practice it could lead to increased calf mortality or increased veterinary fees at calving. The increase in efficiency from crossing to a large sire was greater the greater the number of calves per dam, but unfortunately so also was the decline to be counter-acted. Since the advantage of a large sire breed was small when the number of calvings per dam was small, the real advantage of a large sire breed was to counteract and greatly reduce the decline in efficiency with number of calvings per dam (Table 20.1).

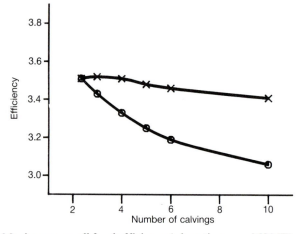

Fig. 20.2. Maximum overall food efficiency (g lean tissue per MJ ME) as a function of number of calvings per dam in a traditional system under best conditions for pure-bred progeny of standard body size (o) and cross-bred progeny of large body size (x).

Table 20.1. Change (%) in maximum overall food efficiency with number of calvings per dam in a traditional system under normal and best conditions with progeny of standard and large body size

Size of sire	Conditions	Number of calvings per dam					
		2/r	3	4	5	6	10
Standard	normal	0	− 2.5	− 6.8	− 10.7	− 12.9	− 17.5
	best	0	− 2.3	− 4.6	− 7.4	− 9.1	− 12.8
Large	normal	0	− 0.7	− 1.8	− 3.6	− 4.6	− 6.8
	best	0	0.3	0.6	− 0.9	− 1.4	− 2.8

Traditional system extended to include bred heifers

A bred-heifer system is taken to be a traditional system in which non-replacement female calves are bred once before slaughter. Under normal conditions in a traditional pure-breeding system, the introduction of bred-heifers was usually mildly disadvantageous, because it required that the dam age at culling be increased (Table 20.2). However, if a large terminal sire was employed as well as bred heifers, overall food efficiency could increase by up to 5 per cent (Fig. 20.3). A further increase in efficiency resulted if the bred heifers were slaughtered somewhat earlier than normal, that is, earlier than 80 per cent mature. At best, with a large terminal sire, early breeding and many calves taken from each dam, bred heifers increased the efficiency of a traditional system by about 10 per cent (Table 20.2). In general, these improvements would be at some extra cost in management.

Table 20.2. Change (%) in maximum overall food efficiency with number of calvings per dam in a traditional system with bred heifers included under normal and best conditions with progeny of standard and large body size

Size of sire	Conditions	Number of calvings per dam					
		2/r	3	4	5	6	10
Standard	normal	0	− 0.4	− 1.1	− 1.8	− 2.5	− 3.9
	best	0	1.1	2.0	1.4	1.1	0.6
Large	normal	0	2.1	5.7	6.4	6.8	7.5
	best	0	4.5	8.0	8.8	9.1	9.9

'All-male' system

Given sex-control, it is usual to suggest that mainly males would be used for beef production and mainly females for dairy production. However, this is not so. Males grow to about 1.4 times the adult weight of females and have leaner carcasses. As individuals, male calves are therefore commercially

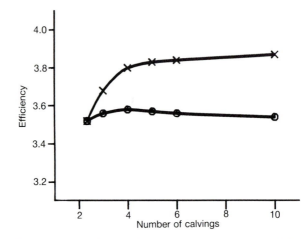

Fig. 20.3. Maximum overall food efficiency (g lean tissue per MJ ME) as a function of number of calvings per dam under best conditions in a traditional system with bred heifers, for purebred progeny of standard body size (o) and crossbred progeny of large body size (x).

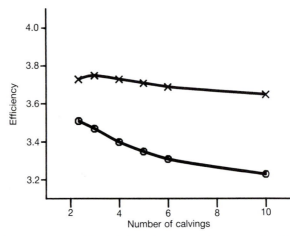

Fig. 20.4. Maximum overall food efficiency (g lean tissue per MJ ME) as a function of number of calvings per dam in an all-male system under best conditions for purebred progeny of standard body size (o) and crossbred progeny of large body size (x).

more valuable than females of the same breeding. However, the body-size advantage of all-male calves would on average be equivalent to cross-bred progeny from a sire breed about one-third larger than the dam breed. Unlike sexed females, sexed male embryos cannot increase food efficiency by reducing maternal overheads, that is, by reducing the number of calvings per dam. Sexed males therefore provide no more than an alternative to cross-breeding.

The extent to which an 'all-male' system declined in efficiency with number of calvings per dam is shown in Fig. 20.4. For pure-bred progeny the decline was steep. With large cross-bred progeny the decline was greatly reduced and an 'all-male' system consistently gave about 7 per cent higher overall efficiency than a traditional system (compare Tables 20.1 and 20.3). Such an advantage, however, would soon be eroded if there was an associated increase in dystocia.

Table 20.3. Change (%) in maximum overall food efficiency with number of calvings per dam in an 'all-male' system under normal and best conditions with progeny of standard and large body size

Size of sire	Conditions	Number of calvings per dam						
		1/r*	2/r	3	4	5	6	10
Standard	normal	—	0	− 1.8	− 4.6	− 7.1	− 9.3	− 12.5
	best	—	0	− 1.1	− 3.1	− 4.6	− 5.4	− 8.0
Large	normal	—	8.2	7.5	6.4	4.3	3.2	1.1
	best	—	6.3	6.8	6.3	5.7	5.1	4.0

* Progeny would be females.

To give a more useful assessment of sex controlled systems, breeding cost was included in the model. An increase in breeding cost quickly led to the 'all-male' system being less efficient. A ten-fold increase in breeding cost reduced 'all-male' efficiency by 9 or 10 per cent. If breeding cost with sex control was more than about five times the cost of a normal insemination, the advantage of an 'all-male' system was lost.

Single-sex bred-heifer system

A main conclusion from the three production systems studied above was that, in general, fewer calvings per dam gave higher efficiency. The minimum number in a traditional herd is on average about 2.4 calvings to yield a replacement female calf. With sex control to ensure the birth of only females, a replacement calf could be produced in 1.2 calvings. The food required to grow progeny to slaughter would remain unchanged, but more than a year's food intake by the dam would be saved and her individual food efficiency would increase sharply; and so, in consequence, would overall efficiency. When a dam is slaughtered shortly after her first calf is weaned, there is a sub-stantial increase in efficiency because the dam herself assumes the role of slaughter offspring and most of the conventional maternal overhead cost of producing a calf disappears by becoming part of productive growth. Apart from about 20 per cent of dams having a second calf, a bred-heifer line is thus an efficient way of producing beef.

Breeding costs   Figure 20.5 illustrates how overall efficiency in an SSBH system changed with breeding cost, from a high efficiency at standard AI cost to very low when 50 times standard cost. Breeding costs are therefore critical in an SSBH system.

Degree of maturity at slaughter   Bred heifers normally calve when about 0.8 mature in body weight. Maiden heifers are normally slaughtered when 0.5 to 0.7 mature. This delay in slaughter of bred heifers carried an efficiency penalty (Table 20.4). When breeding cost for a sexed embryo was 5, 10, 25, or 50 times standard AI cost, optimal degree of maturity at slaughter was about 0.4, 0.5, 0.6, or 0.7. These theoretical optima are unlikely to be attained when breeding cost is less than 25 times standard cost. In practice, optimal slaughter should be as early as attainable (Table 20.4). Early breeding and slaughter clearly become traits of some importance.

SSBH system and breed size   Different breeds and crossbred types are all eligible for bred-heifer production (Crowley 1973). In an SSBH system, there is no dam maintenance requirement and a large heifer calf increases both its input and output proportionally with no change in food efficiency. Hence, the overall food efficiency of an SSBH system is independent of the genetic body size of a bred heifer. Breeds could be chosen to match seasonal or other variations in food supply.

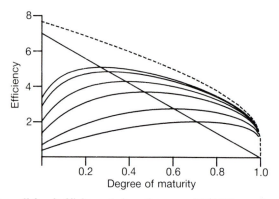

Fig. 20.5.   Overall food efficiency (g lean tissue per MJ ME) as a function of degree of maturity of progeny at slaughter in an SSBH system as breeding costs increase from standard AI cost (top curve) to 2, 3, 10, 25, and 50 times standard cost (bottom curve) under best conditions. Straight line and upper limit curve (---) as for Fig. 20.1.
(Reproduced by courtesy of Animal Production.)

Table 20.4.   Overall food efficiency (E_{bh} in g lean tissue per MJ ME) in an SSBH system with different breeding costs (as multiples of standard AI cost) and slaughter at different degrees of maturity (u_{bh}). Also maximum efficiency and optimal degree of maturity at slaughter. (From Taylor *et al.* 1985.)

u_{bh}	Breeding cost					
	1	2	5	10	25	50
0.5	4.82	4.67	4.28	3.76	2.75	1.90
0.6	4.47	4.36	4.07	3.65	2.80	2.01
0.7	4.04	3.97	3.75	3.43	2.74	2.06
0.8	3.51	3.46	3.31	3.09	2.71	2.01
$(E_{bh})_{max}$	5.29	5.00	4.40	3.77	2.80	2.06
$(u_{bh})_{opt}$	0.24	0.28	0.37	0.46	0.60	0.71

Technological feasibility   Sexed sperm or sexed embryo transfer is a biological prize often claimed but never won. The 1983 headline 'Sexed embryos now possible' was an unfulfilled promise. What will be the fate of the 1985 headline 'World's first embryo to be split and sexed'? More seriously, commercial application of the SSBH concept will be possible with the development of improved technology of sex control as discussed in Kiddy and Hafs (1971), Betteridge *et al.* (1981), and White *et al.* (1982). The best method for producing sexed male embryos might be quite different from the best method for producing sexed female embryos. Both costs and levels of success could be different. For example, a method of killing male sperm or embryos would be ideally suited to an SSBH system.

SSBH system and twinning   How does a single-sex bred-heifer line compare with a twin-producing line in terms of food efficiency? Theoretically, if bred heifers could be slaughtered at an early stage of maturity, all maternal overhead costs are eliminated because the separate roles of dam and growing progeny become merged into one. Twins are efficient because they halve maternal overhead costs. If, at the limit, there were no such costs, halving them gives no further saving, so that bred heifers with twins are no more efficient than bred heifers with singles. Twins and sexed female embryos must therefore be regarded as alternative ways of increasing efficiency by reducing maternal overhead costs.

Comparison of four systems of beef production
When overall efficiencies attainable by different beef production systems were compared (Fig. 20.6 and Table 20.5), the SSBH system clearly gave potentially the highest overall efficiency. By comparison the other three systems appeared quite similar in overall efficiency. For technological

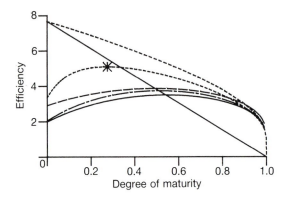

Fig. 20.6. Overall food efficiency as a function of degree of maturity of progeny at slaughter for an SSBH system (---), compared with a traditional system with bred heifers (--), an all-male system (---) and a traditional system (——), all with a large terminal sire and under best conditions. Straight line and * give points of maximum efficiency. Also upper limit curve as in Fig. 20.1. (Reproduced by courtesy of Animal Production.)

efficiency levels given by breeding costs that were ten times standard AI cost, the SSBH system with normal reproductive rates was still slightly more efficient than a traditional system under comparable conditions. With breeding costs in the range 10–15 times standard, the SSBH system was superior only if heifers were bred and slaughtered at a relatively young age, or if a reproductive rate of $r = 1$ was achieved through multiple births. If technological efficiency is not high, both SSBH and 'all-male' systems should be avoided. If semen with a sex ratio in favour of males were available at no extra cost, the 'all-male' system might offer some small improvement without serious risk.

Table 20.5. Superiority (%) of three systems of beef production over a traditional system for maximum overall efficiency with different breeding costs, progeny size and conditions. Optimum number of calvings per dam in brackets

System	Breeding cost	Progeny size	Conditions normal (*n*)	best (*n*)
Traditional with bred heifers	1	1.4	7.5 (10)	9.3 (10)
'All-male'	1	1.4	8.2 (2.7)	6.2 (3)
'All-male'	10	1.4	−2.0 (2)	−5.1 (3)
SSBH	1	—	24.6 (1.3)	26.6 (1.2)
SSBH	10	—	6.4 (1.3)	3.4 (1.2)

Conclusion

Single-sex bred-heifer lines provide the basis for a novel system of beef production which merits consideration because of the high levels of efficiency potentially attainable. High technological efficiency will be required to produce exclusively female embryos. Once operational, however, an SSBH scheme with normal reproductive rates is ideally structured for practising genetic selection. Operating an SSBH system successfully will require a breed type capable of early breeding and of producing a high quality carcass at an early age.

References

Betteridge, K. J., Hare, W. C. D., and Singh, E. L. (1981). Approaches to sex selection in farm animals. In: *New technologies in animal breeding* (eds. B. G. Brackett, G. E. Seidel Jr., and S. M. Seidel) pp. 109–125, Academic Press, New York.

Crowley, J. P. (1973). The facts of once-bred heifer production. In: *The Maiden Female — A Means of Increasing Meat Production* (ed. J. B. Owen) pp. 8–17: Univ. Aberdeen, U.K.

Kiddy, C. A. and Hafs, H. D. (1971). *Sex ratio at birth — prospects for control.* Amer. Soc. Anim. Sci. Albany. New York.

Taylor, St. C. S., Moore, A. J., Thiessen, R. B., and Bailey, C. M. (1985). Food efficiency in traditional and sex controlled systems of beef production. *Anim. Prod.* **40**, 401–40.

Thiessen, R. B., Hnizdo, E., Maxwell, D. A. G., Gibson, D., and Taylor, St. C. S. (1984). Multibreed comparisons of body weight, growth rate and food intake in British cattle. *Anim. Prod.* **38**, 323–40.

White, K. L., Linder, G. M., Anderson, G. B., and Bondurant, R. H. (1982). Survival after transfer of 'sexed' mouse embryos exposed to H–Y antisera. *Theriogenology* **18**, 655–62.

Discussion

The question was raised as to why previous advice had seemed to be so wrong over the optimum number of calves from a beef cow. The explanation was that the overhead cost of the cow purchase price had been shared over several calves and many advisers had omitted to put an efficiency value on the culled cow.

The need for heifers in the single-sex system to be sexually mature at an early age was discussed. It was considered that if the time of conception could be moved back from the normal 80 per cent, to 60 per cent or less of body maturity, then there would be greater gains in efficiency. It was therefore appropriate to select for early sexual maturity of heifers and for carcass suitability at a relatively early age.

The value of twinning was discussed and the conclusion was that twinning would not be nearly so important in a single-sex system as in a traditional system. For calves coming from the dairy herd, the conclusion was not so clear.

CEC Seminar, Edinburgh, June 19–20 1985: List of Participants by Country

Belgium

Dr Y. Bouquet, Faculty Diergeneeskunde, University of Gent, Heidestract 19, 9220 Merelbeke, Gent.

Mr. J. Connell, CEC, 200 Rue de la Loi, B 1049, Brussels.

Professor R. Hanset, Faculty of Veterinary Medicine, University of Liege. Rue de Veterinaires 45. B-1070, Brussels.

Denmark

Dr G.L. Christensen, National Institute of Animal Science, P.O. 39, 8833, Orum Sonderlyng.

Dr. K. Christensen, Department of Animal Genetics, Royal Veterinary and Agricultural University, Bulowsvej 13, DK 1870 Copenhagen V.

Dr T. Liboriussen, National Institute of Animal Science, P.O. Box 39, 8833, Orum Sonderlyng.

Dr P.H. Peterson, Institute of Animal Science, Royal Veterinary and Agricultural University, Rolighedsvej 23, DK-1958, Copenhagen V.

Dr K. Sejrsen, National Institute of Animal Science, P.O. Box 39, 8833, Orum Sonderlyng.

France

Dr R. Lathe, Institut de Chimie Biologique, Faculte de Medecine, 11 Rue Humann, Strasbourg.

Dr J.C. Mercier, CNRZ, INRA, Laboratory of Physiology and Lactation, 78350, Jouy-en-Josas.

Dr P. Sellier, CNRZ, INRA, Quantitative and Applied Genetics, 78350, Jouy-en-Josas.

Germany

Dr G. Brem, Institute for Animal Breeding, Ludw-Maximilians University, Veterinarstr. 13, 8000, Munchen 22.
Dr M. Förster, Institute of Animal Genetics, Technical University of Munich, D8050 Freising, Weihenstephan, Munchen.
Dr E. Müller, Institute for Animal Breeding, University of Hohenheim, Postfach 700562, D-7000 Stuttgart 70.
Professor Dr D. Smidt, Institute of Animal Breeding, Mariensee, 3057, Neustadt 1.

Greece

Dr S. Baronos, Directorate of Veterinary Research, Ministry of Agriculture, Acharnon Str. No. 2, 10176 Athens.

Ireland

Dr J. P. Hanrahan, Agriculture Institute, Belclare Tuam, Co. Galway.

Italy

Professor D. Di Berardino, Institute of Animal Production, Faculty of Agricultural Science, Portici, Napoli.

Netherlands

Dr P. Booman, Institute for Animal Husbandry, 'Schoonoord', P.O. Box 501, 3700 AM Zeist.
Dr E. Kanis, Department of Animal Breeding, Agricultural University, Wageningen.
Dr D. Minkema, Institute for Animal Husbandry, 'Schoonoord', P.O. Box 501, 3700 AM Zeist.
Dr P. Stam, Department of Genetics, Agricultural University, Wageningen.

United Kingdom

Dr R. Lovell-Badge, MRC Mammalian Development Unit, University College, London, 4 Stephenson Way, London NW1 2MG

Local Participants (Edinburgh)

Dr A. L. Archibald
Mr N. Cameron
Dr A. J. Clark
Dr C. S. Haley
Dr J. P. Gibson

Dr P. Simons
Dr P. Sinnett-Smith
Dr C. Smith
Dr St. C. S. Taylor
Mr R. Thompson

Dr R.B. Land   Mr J. Woolliams
AFRC Animal Breeding Research Organization, West Mains Road, Edinburgh EH9 3JQ.
Dr J. Bishop   Dr A. Leigh-Brown
Dr P. Gupta   Dr T. MacKay
Prof. W.G. Hill   Prof A. Robertson
Institute of Animal Genetics, West Mains Road, Edinburgh EH9 3JQ.
Prof. J.W.B. King,
AFRC Animal Breeding Liaison Group, West Mains Road, Edinburgh EH9 3JQ.
Dr J.C. McKay   Dr H. Sang
Dr M. Perry   Mr A. Tinch
AFRC Poultry Research Centre, Roslin, Midlothian EH25 9PS.

Index

197

DATE DUE

~~NOV~~			

DEMCO 38-297